# THE ORGANIC ARTIST

# THE ORGANIC ARTIST

Make your own **PAINT, PAPER, PENS, PIGMENTS, PRINTS,** and more from **NATURE**

## NICK NEDDO

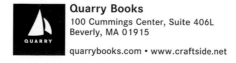

**Quarry Books**
100 Cummings Center, Suite 406L
Beverly, MA 01915

quarrybooks.com • www.craftside.net

First published in the United States of America by
Quarry Books, a member of
Quarto Publishing Group USA Inc.
100 Cummings Center
Suite 406-L
Beverly, Massachusetts 01915-6101
Telephone: (978) 282-9590
Fax: (978) 283-2742
www.quarrybooks.com
Visit www.craftside.net for a behind-the-scenes peek at our crafty world!

Library of Congress Cataloging-in-Publication Data

Neddo, Nick.
  The organic artist : make your own paint, paper, pens, pigments, prints, and more from nature / Nick Neddo.
     pages cm
  ISBN 978-1-59253-926-0 (paperback)
  1. Artists' materials--Formulae. 2. Natural products. 3. Nature craft. I. Title.
  N8530.N38 2015
  702.8'4--dc23
                          2014025582

ISBN: 978-1-59253-926-0

Digital edition published in 2015
eISBN: 978-1-62788-225-5


10 9 8 7 6 5 4 3 2 1

Cover and Book Design: Raquel Joya
Illustrations: Nicholas Neddo
Photography: Susan Teare
Chauvet Cave images on pages 16 and 137 courtesy of Dr. Jean Clottes

Printed in China

This book is dedicated to the sacred dance of the living world
around us and the Muse—that seductive spirit of creative genius.
May we all welcome her for everyone's sake.

# CONTENTS

# INTRODUCTION

## WHO IS THIS BOOK FOR?

This book is for artists, naturalists, doers, makers, crafters, enthusiastic hobbyists, and creative thrill seekers.

This book is more or less organized like a cookbook. Each chapter has a series of projects with step-by-step instructions for you to follow to make your own tool or material. The illustrations are intended to demonstrate the medium being featured in each section of the chapter, and in some cases, pay respect to the species from which the specific medium is derived. Although each chapter topic can be linked to topics from other chapters, they stand alone and can be read in any sequence.

## PUTTING THINGS INTO CONTEXT

Consider for a moment the modern human timeline on Earth. It's widely accepted by archaeologists that *Homo sapiens* have been on the scene for at least 200,000 years. Of that 200,000 years, we have been living in a civilized state for only 10,000 years. This means that for more than 95 percent of our time, we lived with stone tool technology. Ironically, the vast majority of time that our species has been alive is referred to as prehistory.

Let's give our ancestors the credit they deserve. If not for their creative ingenuity, we would not be here to read these words. Their intimate knowledge of the landscape and natural world at large gave them everything they needed to build their entire material culture by hand. For perspective, think about how many things we use each day in our lives that we do not know how to make from the landscape by hand.

This book is not strictly coming from specific cultures or technologies. Much of its inspiration, however, is directly linked to a lifelong pursuit of trying to see the world from a problem-solving perspective, with a wilderness technology lens, based on ever-growing intimacy with the living world we call the landscape.

Bamboo is a versatile material, with countless applications in the realm of making things. Bamboo Forest.

# WHY MAKE YOUR OWN ART SUPPLIES?

Making your own tools (and processing materials for doing so) from the landscape is unbelievably satisfying on a profound and even instinctive level. Much of this satisfaction comes from the process of transformation that occurs each time we make something from another thing. One of the results of making things from the landscape by hand is the unavoidable deepening of one's knowledge of (and relationship to) the local bioregion where we live. Through working with raw materials, we begin to learn to speak the language of that particular material. We have to use our awareness to observe the specific characteristics, strengths, and limitations that are unique to the material. Through this level of interaction a conversation begins. We learn to be receptive to the feedback the raw materials provide as we manipulate them to take the shape and function that we desire. Ultimately this level of participation with the landscape is a path to help us remember that we are part of its natural history and ecology, not just a visitor like an astronaut on a foreign planet.

# ABOUT THE ARTWORK

I created all of the illustrations you see within the pages of this book with art supplies, tools, and materials that I made myself. My intention with the subject matter within most of the compositions is to represent the creatures that provided the raw materials to make the medium in which they are depicted. This is one reason you see so many trees in this book. When using ink I made from black walnut trees, with a pen I made from a black walnut twig, I am compelled to make images of this tree. A crayon drawing of bees on honeycomb is a direct homage to the beeswax I used as a base for the crayons. Other pieces are less intertwined in this way, but I included them to show examples of artwork made with the relevant tools and materials represented in the chapter, with their uniquely rustic and surprisingly elegant characteristics.

I made each project featured as well. I aspire to make tools that are beautiful in their function and that retain as much of their raw form as possible.

*Small Celebration of Red Oak*. Nick Neddo 2013, pen, brush and acorn ink. The bark, leaves, twigs, and acorns of the red oak tree, the source of the ink that was used to make the image.

# ACCESSING THE LANDSCAPE

The landscape is all around us, whether you live in the countryside, suburbs, or city. The raw materials of the landscape will vary, depending on the location and surroundings. This book focuses on the parts of the landscape that are expressing an ecosystem that is somewhere in the stages of that place's natural succession. In other words, it focuses on places that are allowed to grow and develop as they choose, without the constant hand of humankind's obsession with control and notions of tidiness. These are the larger, unbroken wildernesses as well as the miniature forgotten wildernesses that exist behind the gas stations and shopping malls, between the houses and roadways, and along the waterways from great rivers to diminutive trickling streams.

Many people do not experience these natural treasures because they are afraid of some mysterious possible hazard. People tend to fear things that they do not understand, and we destroy the things that we fear. Does this insight have any relevance in the human role of today's ever-threatening environmental collapse?

Let us replace fear of the unknown with curiosity. When we begin to learn new things about a potentially dangerous entity, we become empowered with knowledge, rather than limited by fear and ignorance.

Learning about the natural history of the places we call home is the underlying, baseline priority if we want access to the mysteries, lessons, and beauty of the living world around us. Begin by getting some field guides that can help you learn to perceive nature in a different way. Most importantly, spend time outside with the landscape. Intentionally interact with different critters and entities. Are you curious about a particular plant? Sit down with it and hang out. Make some observations. Let your curiosity guide how you interact with the landscape. Be authentic with your pursuit of knowledge and relationship with the landscape you live with. Do not let others define everything for you. Direct interaction with the other species in the bioregion we inhabit leads us to remember dormant parts of our selves. We learn to measure quality of life beyond material wealth. We become participants in the timeless tradition of give and take as we harvest and tend to the health of plant and animal communities.

This way of being on the earth creates an ever-growing and deepening belonging to and love for the living world around us. This is the key to providing for your material needs directly from the landscape.

## USING RAW MATERIALS

Each species that provides the raw materials for the projects in this book has its own personality, and therefore requires the artist to slow down enough to observe these subtle nuances. If we try to force the material to our will, often the project will fail. However, when you begin to learn the limitations and allowances of the material, a conversation of sorts begins. This is when the creative process becomes a conscious interaction with the creatures that created the materials we hold in our hands. This conversation leads to insights that otherwise go unexperienced and an awareness of complex and beautiful relationships that otherwise go unacknowledged.

## HARVESTING FROM THE LANDSCAPE

The ecological consequences of our behavior on the landscape can vary quite dramatically based on how we approach the task. We can take in ways that cause harm, where the land (and biodiversity) suffers a net loss, or we can take in ways that allow the land to favor greater biodiversity, causing a net gain. This all has to do with our awareness, education, and often our intuition.

Answer the following questions before taking a life whether it be an animal or a plant:

- Is it necessary to kill this being to make what I want to make? Sometimes it is and sometimes it is not.

- If so, is this the right one of its kind to take? I try to answer this question by looking for opportunities to improve the lives of those in immediate proximity to the creature in question.

- For example, is this tree/shrub engaged in direct competition for resources with its neighbors? If trees are growing too close together, neither of them will get enough of what they need (sunlight, soil nutrients, and water) to reach maturity. Often in this situation, both will die before they can reproduce. This gives the participating *Homo sapiens* an opportunity to promote the health and survival of one by harvesting and using the other.

## BEING SAFE AND SAFE TOOL USE

This must be said: Life can be dangerous, and potential hazards are everywhere. There are innumerable ways in which someone can be injured when walking on the landscape, using tools, and making things. By using this book, you, the reader, take full responsibility of your safety. If you are uncomfortable, or feel unsafe at any moment, stop what you are doing and reassess your approach.

The use of tools is a central element of the process of making things. One of the limitations that our species has is our dependence on tools to do most of the things we do in life. Another way to see it is that tool making, as well as problem solving, is one of our "super powers." Our tools become an extension of our bodies, allowing us to do things that were not possible before, and to live places that were previously uninhabitable to us. In essence, the more tools we know how to use (and make), the more options we have in our daily lives.

Tools can be dangerous if they are not used with care and respect. This book is not intended to teach the reader how to use basic hand tools. A baseline level of experience of hand tool use will be invaluable to the reader in the pursuit of the projects that follow on the pages ahead. However, I include some tips here and there for safe and effective techniques for some of the tools mentioned. Look for this information in the tips and sidebars sections in each chapter. Keep in mind that any tool can be dangerous depending on how it is being used. Hammers can smash thumbs as well as drive nails. Life is dangerous, and you get to make the decision as to how you pursue its adventures and rewards. You may want to reconsider pursuing the projects in this book, and stay inside where it is safe, watching reality television instead. That, however, presents different hazards, now doesn't it?

# SETTING UP A STUDIO WORKSHOP

As with many creative ventures, having the space to do your work is critical. There is a spectrum of "needs" to be met to achieve the ideal working environment, but many of these details fall under the "ideal" category and may best be aspired toward later, once you have set up your basic workshop space.

The first thing to figure out is where you can make a mess and keep it open while you are doing other things. When we feel inspired to create (whether it's a painting or carving a wooden spoon), it's important to have an infrastructure in place that allows the creative process to begin (or continue) with as few logistical obstacles in the way as possible. Find a space you can claim as a studio/workshop. You may be fortunate enough to designate a separate building or room for this, or you may have to make do with a desk in the corner of a room, or even a portion of the kitchen table. (Try to avoid this at all costs.) Wherever your studio/workshop is, protect and maintain it as a sacred center for your creative miracles. You may find that your quality of life improves by making this one adjustment. Convince your spouse and housemates that your having a creative workspace will make their lives better, too.

In this space, you can keep your tools and materials, make your messes, explore crazy notions, and ultimately give yourself permission to be the eccentric creative that you are.

Humans need to stay connected to creativity in order to access our problem-solving prowess, so in all seriousness, the world may depend on your having this space.

# EXPERIMENTATION AND PLAYFULNESS

This book was born out of a collection of loves and infatuations. The foundation is that of the living world we live with and participate in. Another is the love of art and the creative process in general. Yet another love is that of Stone Age living skills or primitive technology. Many of the projects presented are original ideas that came about from a challenge: How can I make one of these without using metal, plastic, or glass? What characteristics am I looking for in this element of this tool? What creatures have these traits and can help me build this tool?

Give yourself permission to play and pursue your creative whims. This is how original materials, concepts, and art come into the world. The key is to have the courage to make mistakes and be confronted with failure. Failure is the result of hitting a roadblock along the way and letting it stop

the exploration and thus preventing a lesson from being learned. Failure occurs when the lesson is not pursued as a sacred element to the art of figuring stuff out.

So, dear reader, I would like to outline my intentions for you here. Take on an adventurous spirit and give your creativity the right of way when it bubbles up in you; approach nature with curiosity and respect, and let your intuitions and creative whims guide you on your journey; and venture out on the landscape as a participant, exploring and experiencing the forgotten wilderness where you live and visit.

Ultimately, the intent of this book is to help strengthen the connections we all have to the living world around us.

# 1

# CHARCOAL

Charcoal, the carbon remains of burned objects, is perhaps the artistic medium that has been used longer than any other. Charcoal is found on cave walls, along with other pigments, that date back more than 30,000 years.

The oldest known examples of charcoal in ancient cave art (parietal art) adorn the walls of Chauvet Cave in France. Those Stone Age artists used charred sticks to make bold lines as well as fine ones, and they used an entire grayscale of tones for exquisitely rendered shading.

Using charcoal brings a primal satisfaction to the creative experience. Because charcoal is relatively soft in density, it allows the artist to feel the texture of the surface (the support) underneath in a way that is unique to this medium. Charcoal allows you to cover and shade large areas quickly, and thus it has traditionally been used as a medium for sketching and making drafts of more sophisticated works. However, charcoal can certainly stand alone as its own medium; the creative possibilities are only limited by the imagination of the user.

An assortment of various homemade charcoal drawing tools and materials

# Simple Charcoal

Simple charcoal is basically the black, partially-burned pieces of wood scavenged from the previous night's camp fire. This, most likely, was the first form of charcoal that our ancient ancestors used to represent the world around them on cave walls and other long-gone surfaces.

## MATERIALS

Various sizes of firewood: tinder, kindling, sticks

## TOOLS

Fire starting device (matches, lighter, bow drill, etc.)

## 1: Make a campfire.

Find a responsible place to make a small campfire, and be sure to use your common sense with regard to safety. Begin by collecting firewood. Make sure that you do not try to gather wood from green (living) branches. This will damage healthy trees, make for a frustrating fire-building experience, and make you look like a fool to any observers. Instead, look for dead branches in the understory of living trees, or find trees that are dead altogether. The driest fuel wood is found on standing dead trees, although you can often get sufficiently dry wood from the ground, depending on the local climate and recent weather conditions.

Bring the firewood back to the fire pit and break it into smaller lengths (around 12 to 16 inches [30.5 to 53.5 cm]), arranging it in three piles of increasing diameter. Next, take a bundle of the smallest wood (these twigs should be the diameter of pencil graphite) and bend/break it in half, making a miniature A-frame in the center of the fire pit. Now, place the next size up (about pencil diameter) at a 45° angle around the tiny twigs, making a mini tipi-like structure. Continue this pattern, adding the next size up (about the diameter of a marker), evenly spaced around the circumference of the fire structure. When you have built the fire structure, light it, sit back, and enjoy its companionship.

## 2: Collect the charcoal fragments.

When the fire has burned out and cooled off, scavenge the remains to find the black charcoal fragments that are left behind. This is essentially the material that our ancient ancestors first used for decorating the walls of caves.

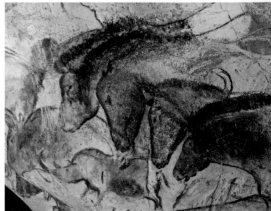

A hearth for manufacturing charcoal drawing sticks (top) and some of the earliest known human artwork (bottom), both dated nearly 32,000 years ago. These were both discovered in the legendary Chauvet Cave in southern France. Notice the sophisticated use of charcoal in the painting.

# Charcoal Sticks

This is the type of charcoal that you can buy in art supply stores. These sticks of charcoal are traditionally made from wild grape vine and willow, although you can use any species of wood, experimenting with various densities to get variations in the hardness. To make charcoal sticks, you first need to make a simple tool called a charcoal kiln. This is a little metal vessel that can be loaded with sticks and twigs, sealed shut, and then placed in a hot campfire to char. Charcoal kilns allow the contained materials to burn without entirely combusting, due to the low oxygen environment inside the kiln.

Finished charcoal sticks in the kiln

## MATERIALS

Grape vine
Willow rods
Firewood

## TOOLS

Sharp knife
Hand pruners
Tin with a tightly fitted lid
    or new, empty paint can
    with a lid
Fire

**TIP:** Check your local flea market or thrift store for used cookie tins.

Tools used in this chapter: metal container with tight lid, fire, folding saw, hand pruners, and pocket knife

## 1: Find a potential charcoal kiln.

Look for a tin container with a tightly fitted lid, such as a cookie tin or a new, empty paint can from a paint or hardware store. Be sure that it is all metal.

**TIP:** You can test multiple tins at once.

## 2: Season/test the tin.

Secure the lid to the tin. Make a campfire structure and place the tin in it before igniting it. Light the fire and allow the tin to be engulfed in flames as the fire burns. When the fire is out and the tin has cooled off, it is safe to handle the tin to see how well it held up.

Not all tins are created equally. Some tins will become distorted and disfigured as a result of this trial. These are not cut out for the task of making charcoal sticks. Due to warping and melting, the lids on some tins will lose their fit, and any attempts to make charcoal sticks in them will be counter-productive. It is better to learn whether or not your tin is suitable for the role before you put your sticks into them. Failure of the charcoal kiln (the tin) most often results in the loss of your processed material. If the tin fails, find another one and repeat this test. If, however, your little tin survives the ordeal and maintains a solid relationship with its lid, you can move on to the next step, which is making charcoal sticks in it.

## 3: Find some sticks.

Now that you have a charcoal kiln, you can begin your hunt for some sticks. Like I said earlier, these can be any sticks (that are not poisonous), but I will suggest wild grape vine (*Vitis* spp.) and any species of willow (*Salix* spp.).

Look for wild grape along overgrown edges between fields, near forest edges, and along waterways. Willows are water-loving, and they are reliably found near wetlands, streams, and rivers. Spend some time getting acquainted with these critters (in a field guide or in some other way) before you go out cutting things down. Take what you need for your project and leave the rest for the wildlife, which may include you one day if you keep doing this kind of stuff.

## 4: Remove the bark.

The finest charcoal sticks are made from twigs that are debarked. With the bark removed you will have a consistent density throughout the piece of charcoal, thus giving the artist more control and predictability of the charcoal's function. Because the bark has a different density than the wood, it is important to remove it before progressing from here. If the bark is left on the twigs, the resulting charcoal sticks will be of lower quality due to the outside of the stick behaving differently than the inside. Use the dull edge on the back of a pocket knife to scrape the bark off.

**TIP:** Set aside the bark that you remove as a potential fiber source to use to create pulp for papermaking. (See "Paper" on page 123.)

## 5: Trim and pack.

Use a set of hand pruners to cut each stick to a length relative to the size of the height of the charcoal kiln. (Make sure they fit.) Use one stick as a guide to cut the other sticks to the same length. Then pack the kiln as tightly with the sticks as you can while still being able to close the lid. Punch a small hole in the lid with a nail and hammer to let the gasses out during firing.

Raw wild grape (*Vitis* spp.) vine and willow (*Salix* spp.) sticks, ready to be processed into charcoal

**3**

Debark the sticks with a scraping tool, such as the dull edge on the back of a pocket knife. This step ensures consistent density and reliability of your charcoal sticks.

**4**

**TIP:** The sticks you put into the tin will shrink significantly during their transformation into charcoal. Anticipate this when you cut your twigs and sticks to length, making them as long as possible while still being able to fit into the kiln with the lid fully closed. Also, consider this shrinking in regards to the diameter desired for your finished charcoal sticks. Choose sticks of various diameters to ensure that you get what you want.

**5**

Snip the debarked sticks to fit snugly into the kiln. They will shrink in length and diameter as they fire, so pack them in tight and keep them as long as possible while ensuring the lid makes a secure fit.

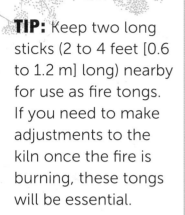

**TIP:** Keep two long sticks (2 to 4 feet [0.6 to 1.2 m] long) nearby for use as fire tongs. If you need to make adjustments to the kiln once the fire is burning, these tongs will be essential.

## 6: Secure the lid.

As a protective measure, you may want to lock the lid down with some metal wire in case of distortion or other potential stress that may compromise the kiln's seal during firing. A simple, single wrap with thick-gage metal wire, tied down with a pair of pliers, will do the job well. Another option for keeping the lid on the kiln while firing is to place a weight, such as a rock or brick, onto the top. The consistent downward pressure will keep the lid from popping off in the fire.

## 7: Fire the charcoal.

With the lid locked on and your sticks inside, you are ready to char them. Make a campfire structure and place the loaded tin inside. Place a new round of firewood on top of, and around the tin and then ignite it. Add more firewood as needed to allow it to get really hot for about an hour, burning the sticks inside. Enjoy the fire and when you are ready to move on, let it burn out.

## 8: Remove the sticks.

When the fire is dead and the kiln is cool, remove the kiln. Open the lid to find your fresh batch of handmade artist charcoal sticks.

## Using Charcoal Sticks

One distinctive characteristic of charcoal as a creative medium is its ability to smudge easily. This trait can be used to produce impressive subtleties of tone, but it also allows finished pieces to be damaged by accidental smudging. You can easily solve this problem by applying a fixative, which protects the surface from smudges and smears.

**7**

Build a fire with the kiln incorporated into the structure. Once lit, tend the fire by surrounding the kiln with firewood so it is exposed to heat from all sides. Make sure the lid stays on throughout this process!

**8**

Finished charcoal sticks in the kiln

*Small Celebration of Willow.* Nick Neddo 2013,
charcoal drawing with charcoal made from willow

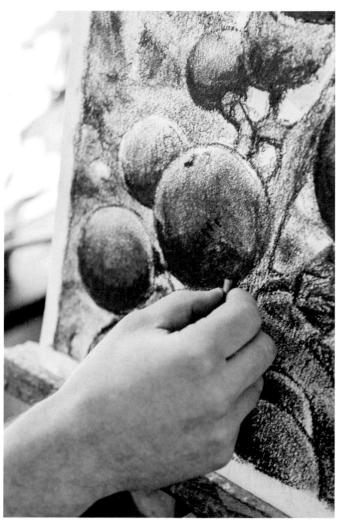

The author demonstrates the use of
charcoal as a drawing medium.

**STORAGE TIP:** Charcoal sticks are fragile, and they should be kept in
a safe place where they will not be crushed or bent. Find a rigid-walled
vessel, such as a glass jar or tin can, to keep them in. Keep your eyes
out for small boxes or other containers to use as protective storage
when you need to transport the charcoal sticks.

# Charcoal Holders

Charcoal can be messy, and that is part of the fun sometimes. Other times you may not want to have all of that extra pigment all over your hands, leaving grubby little fingerprints all over your work. This is when a charcoal holder comes in handy. These tools are relatively simple to make, and they essentially add an extension to your charcoal stick, increasing the length of your tool and giving you a non-messy surface to grip.

Charcoal holders are also helpful when your hand might get in the way, inadvertently rubbing across your work. The extra length of the charcoal holder allows the artist to relax the hand and achieve longer, more consistent strokes.

You will probably want to make several holders, each of varying dimensions to accommodate a variety of charcoal stick diameters.

An assortment of charcoal holders made from several species including bamboo (*Bambuseae*), black walnut (*Juglans nigra*), elderberry (*Sambucus* spp.), honeysuckle (*Lonicera* spp.), blackberry (*Rhubus allegheniensis*), and staghorn sumac (*Rhus typhina*)

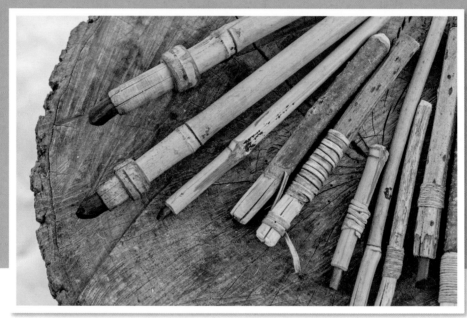

## MATERIALS

Hollow (or pithy) stick, such as bamboo, elderberry, sumac, or blackberry
String (optional)

## TOOLS

Folding saw
Sharp knife
Awl or improvised reaming tool, such as a small, straight, firm stick
Light mallet or any small stick
Sandpaper (optional)

## 1: Find sticks.

You'll need to consult your naturalist knowledge or field guides to identify shrub and tree species that grow sticks that are either hollow or pithy (foamy or corky in texture) inside. These critters are well suited for making charcoal holders, as well as pens and paintbrushes. Good options include, but are not limited to, walnut trees, bamboo, elderberry, sumac (please, not poison sumac!), and honeysuckle.

## 2: Cut sticks to length.

The length that you cut these sticks is entirely up to you, based on how long you want your charcoal holder to be. I like mine to be somewhere between the length of a pencil and that of a paintbrush (4 to 10 inches [10 to 25.5 cm]). Make one cut 2 inches (5 cm) from a branch node, leaving the node within the length of your holder. Use sandpaper to smooth any rough edges that you have issues with.

Hollow or pithy (soft-centered) sticks of various species suitable for making charcoal holders

### 3: Ream the core.

Using an awl or improvised reaming tool, compress the pithy core in toward the center of the stick about 2 inches (5 cm) deep. This will create a more open center, creating room for the charcoal.

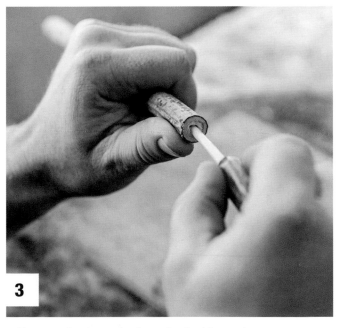

Use an awl or improvised reaming tool to create a hollow in the center of pithy sticks.

## 4: Split the end.

Take the end that is 2 inches (5 cm) from a branch node and bisect the diameter with your knife. With the knife in place, carefully tap the top of the blade with a light mallet (or any small stick) until a split begins to form. Be careful not to force this split beyond the branch node or your holder may end up in two pieces. Now make another split perpendicular to the first one to make an X.

## Using the Charcoal Holder

Once you've made the charcoal holder, it's very simple to use.

Select charcoal sticks that have similar diameter to the opening of your charcoal holder. Break the charcoal stick into a 2- to 3-inch (5 to 7.5 cm) segment and insert one end into the charcoal holder about 1 inch (2.5 cm) deep. This will provide enough surface area contact inside the holder to keep your charcoal stick fitted securely.

Hold the knife with one hand, placing the blade exactly where you want the split to be. With the other hand, lightly tap the back of the blade with a small mallet or stick until a short split forms.

**TIP:** If your charcoal stick breaks while it is fitted into your holder, it can be difficult to grip the stub end well enough to remove it. When pieces get stuck in the holder like this, you can crush the remaining segment of charcoal with an awl or a firm stick of slightly smaller diameter. This will powder the remains of the charcoal stick, allowing you to pour out the residue. Repeat if necessary until the cavity is empty and ready to accept a new charcoal stick.

Charcoal drawing sticks are easy to make and fun to use. The author working on a piece: *Smiling Lady*. Nick Neddo 2013, willow charcoal on paper.

## The Knotless Lash

Lashes are useful for a wide variety of applications, notably attaching two things together, preventing splits in wood from extending, and making comfort grips on pens and other drawing tools. If you continue to make things by hand, this lashing technique will become like an old friend.

First, cut some string or other cordage long enough to provide adequate coverage of the item you want to wrap. Make a bend at one end, leaving a tail several inches (cm) longer than the overall length of the planned wrap. For example: If you want to make a retention lash on a charcoal holder and you plan for it to span 2 inches (5 cm) of length, than make the tail at least 4 to 8 inches (10 to 20.5 cm) long. Align the tail parallel to the rest of the string and lay it in line with the stick.

Next, grasp the long end and begin wrapping over both the short end and the stick, leaving an inch or two (2.5 to 5 cm) of overhang on the short end. Continue wrapping around the stick and over the short end of the string until you get close to the loop that you made when you bent the string to begin with.

Then, thread the end of the strand you wrapped with through the loop. At this point, each end of the string should be extending from the beginning and end of the wrap.

Finally, tug on both ends of the string with opposing tension in line with the stick. The loop will get pulled under the wrap, taking the loose end of the string with it. Pull on the short end until the loop is buried somewhere in the middle of the wrap. Trim off the loose ends, and you're done!

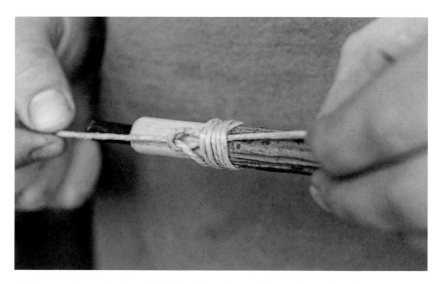

You can add a retention lash if the stick does not have a branch node to control how far the splits extend. In this case just wrap and knot the stick with string about 1 to 2 inches (2.5 to 5 cm) from the end.

# Blending Stumps (Tortillon)

Blending stumps are used for mixing charcoal tones after they've been applied to the paper or other support. These simple tools help the artist take advantage of charcoal's tendency to smudge, greatly expanding the creative options available and significantly affecting the appearance of charcoal on the surface. Fingers work well for this, but they deposit oils, and they're limited in their ability to focus in on the fine details due to their bulkiness and relatively blunt tips. If you're not satisfied with the amount of control you get from your fingertips, you can make blending stumps easily from paper.

The blending stump is used to modulate tones and create subtle changes in value.

## MATERIALS

1 sheet of absorbent paper (approximately 8½ x 11 inches [21.5 x 28 cm]), such as paper towel, newsprint, or your own handmade paper (page 130)
Masking tape or string

## TOOLS

Thin wire or straightened paper clip
Sharp pocket knife (optional)
Sandpaper (optional)
Scissors (optional)

**1:** Fold the paper.

Fold the sheet of paper in half from one orientation, and then fold it again perpendicular to the first fold.

**2:** Roll the paper into a stick.

Find a slender straight object such as a thin wire or straightened paper clip and roll the paper tightly around it at a slight angle. This can be done by intentionally making sure that the straight edges of the paper are not in alignment when you begin the roll. It may take a few tries to get the paper to roll as tight and with the finest tip as possible. Keep trying until you get it the way you want.

**3:** Secure the roll.

Once you have made the roll as tight as you can, with as fine a tip as you desire, lock it together by wrapping the mid-section with a piece of masking tape or binding it with string.

## 4: Remove the wire.

Pull out the wire or paper clip core and then use the scissors to trim your new blending stump to a comfortable length, if necessary. At this point you can snip the ends to make sharp or blunt tips. Now it is ready to use. Make blending stumps with different kinds of paper to experiment with a variety of smudging and blending effects.

**TIP:** You can modify the tip of your stump by carving it to a new point with a pocket knife, or by abrading it to the desired shape with sandpaper. This is often necessary when you need to clean excess pigment from the tip of the stump.

*Small Celebration of Wild Grape*. Nick Neddo 2013, charcoal drawing with charcoal made from wild grape vine

# 2 PENS

In this modern world of abundance and technology, it's easy to take simple tools for granted. Ballpoint and felt-tipped pens and markers are good examples of tools that I had grown accustomed to using.

I must admit that at first the thought of using a dip pen and inkwell seemed inconvenient and cumbersome. I thought, "Why would I want to paint with a stick? That's what paintbrushes are for."

Ultimately, my desire to use my own handmade pens has allowed me to overcome my preconceived notions and open up a world of creative possibility that did not exist with my previously beloved factory pens. Handmade dip pens offer the artist an exciting diversity of line qualities, each specific to the individual pen and the person using it. You can make thin, wispy lines as well as thick, bold lines from the same pen, whereas modern pens have limited capabilities in this regard.

Another unique feature of using dip pens is the ability to control line density and tone. You can take advantage of the varying tones that are presented as the ink flow changes along the spectrum of a fully inked pen, to one that is ready to dip again. You can also play with tone density by diluting the ink.

I still use modern pens in my artwork, but I must say that after using pens made from feathers, sticks, bamboo, antlers, and even bones, my preferences have changed. The former seems sterile and generic, whereas the latter seems to be alive. Each handmade pen has its own unique personality and characteristics. From this perspective, I have to redefine the medium of works created with modern pens as "ink pen," because for me "pen and ink" has become an entirely different medium.

An assortment of pens made by the author

# Simple Pens

Pens don't have to be complex in their design and manufacture. Sticks of various species, sizes, and dimensions can be quite satisfying when used as drawing implements. You can achieve a diversity of effects in line quality when experimenting with sticks and other found objects. These can be slightly modified or used as is.

## MATERIALS

Sticks

## TOOLS

Sharp pocket knife
Glass jar

Sticks of all kinds can be made into simple pens for drawing with ink. Experiment with materials that are local to the bioregion you live in.

## 1: Scavenger hunt.

Go out onto the landscape and find a variety of interesting, pen-size sticks. These can be just about anything as long as they are not poisonous, rotten, or weak.

## 2: Carve the point.

Use a sharp pocket knife to carve one end of the stick to a sharp point. Now remove the bark 2 inches (5 cm) up the pen shaft from the point.

## 3: Prime the pen.

Soak the debarked ends of your pen sticks in a jar of water for 30 minutes or so before using them. This will open the pores of the wood and allow for better ink retention and flow.

The author demonstrates the use of a simple stick pen made from a red oak twig.

Tools used in this chapter: pocket knife, sandpaper, utility knife, awl, candle, paperclip, folding saw, hot plate, and scissors.

# Bamboo (Reed) Pens

Reed pens are believed to be the earliest type of pen, first used by the ancient Egyptians followed by the Romans. They were in wide use for writing and drawing up until the nineteenth century, when steel nib pens became popular.

In general, reed pens are somewhat rigid, with thick and fibrous nibs. The line character of reed pens is directly related to the woody material's lack of supple responsiveness. They seem to have a raw, almost harsh quality that you can use to create striking effects in your drawing.

Bamboo (*Bambuseae*), common reeds, and other woody materials with hollow or pithy cores work well for making pens.

Of the many possibilities to choose from that meet the hollow or pithy criteria, bamboo is borderline perfect as a material choice, due to its relative strength and flexibility. Bamboo nibs that are crafted to a fine point tend to keep their integrity longer than that of quill pens, which need to be resharpened more often. The process of making pens from bamboo or other woods is similar to that of making quill pens from feathers. (See Quill Pens on page 36.) The differences are subtle and are distinctive to how the raw material at hand wants to be worked.

A collection of bamboo pens, each with unique nib characteristics and dimensions for specific line properties

## MATERIALS

Bamboo stick of approximately ¼ to ½ inch (6 mm to 1.3 cm) in diameter and 8 to 10 inches (20.5 to 25.5 cm) long

## TOOLS

Sharp knife or fine-toothed saw (such as a hacksaw)
Awl, nail, or sharp, thin twig
Small mallet (optional)
Cutting board (optional)
Sandpaper

**TIP:** If you want to make pens from solid woods, you can drill out the centers with a power drill. This requires a clamp to hold the pen blank while you drill into it. It can be tricky to get the hollow perfectly centered, but practice and experience will build accuracy. Once you have succeeded in making some sticks hollow, you can proceed with the steps as you would with bamboo.

*Bamboo Forest*. Nick Neddo 2013, bamboo pen and pinesoot ink drawing

Bamboo sticks ready to be made into pens

## 1: Get some bamboo.

Bamboo grows *almost* all over the world, and it's one of the fastest growing plants alive. If it grows where you grow, give yourself a moment to embark upon a scavenger hunt in which the only item on the list is a length of bamboo of pen-like dimensions. You might as well be proactive and bring home a little extra to practice on and get the feel for the material, and in case your first few attempts do not go as planned. Use a sharp knife or fine-toothed saw (such as a hacksaw) to harvest the bamboo shafts. When making cuts along the length of the stalk, choose areas 2 to 3 inches (5 to 7.5 cm) on either side of a node. The branch nodes are important features to these stalks that you will want to take advantage of later.

Plan B: If you cannot find bamboo growing on the landscape around you, there's a good chance that you may live outside of its range. If you have come to this conclusion and are not interested in working with imported materials, you can proceed with reeds, staghorn sumac, elderberry, black walnut, or some other hollow woody material. If, however, you still want to experience the joys of working with bamboo for pen-making, you can find it on the "commercial landscape," at your local hardware store or farm and garden supply store, sold as garden stakes. If you have the option, choose the natural tan-colored bamboo instead of the dyed stuff. Many commercial dyes may not be safe to handle regularly.

## 2: Cut to length.

Bamboo grows with segments divided by branch nodes roughly every 8 to 12 inches (20.5 to 30.5 cm) apart. It is these branch nodes that you want to focus on for where you make your cuts. Each pen blank should have at least one node to ensure its structural integrity. Using a sharp knife or fine-toothed saw, cut the bamboo about 2 inches (5 cm) or so beyond the branch node, designating this as the nib end of the pen.

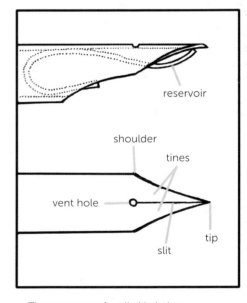

The anatomy of a nib. Variations in the width of the tip are made according to the line qualities desired. Notice how the reservoir is oriented in the pen shaft.

### 3: Cut the first bevel.

Carve the bamboo 1¼ inches (3 cm) from the end that has the node, and to the center of the stalk's diameter (halfway through the cross-section). This cut should be sloping as it begins and then level off quickly once you have made it to the center.

### 4: Clean the inside of the stalk.

Use an awl or even a nail or sharp, thin twig to ream out the pithy material in the center of the bamboo stalk. The area between the tip of the pen and the branch node should be relatively free of any loose material that could possibly interfere with ink flow or reservoir installment. (See Upgrades on page 40.)

### 5: Make the shoulder cut.

The next cut required to make the nib is started ¾ inch (2 cm) from the tip of the pen (in the middle of the span of the first bevel). This cut is also made at a diagonal angle and then levels out once it reaches the depth of the far wall of the stalk. At this stage, the tip of the nib begins to take shape as the width at the end of the stalk narrows a bit.

### 6: Split the center slit.

The center slit is important for facilitating ink flow. Find the middle of the nib tip (from side to side) and place the edge of a sharp knife there. With your thumb on the back of the blade, apply slight pressure toward the node, while gently rocking back and forth until the split forms. The branch node will prevent the split from extending too far along the length of the pen. If you prefer, you can use a small mallet to lightly tap the back of the knife blade to make the slit.

### 7: Shape the tines.

With the center slit in place, you can trim the tines of the nib to be symmetrical to one another, with the slit being the midline. Make the nib according to your preference: narrow for drawing thin lines or wide for making bold lines.

### 8: Cut the nib angle.

This cut is made from the outer wall side of the stalk, toward the inside. Place the nib tip onto a cutting board or other suitable surface that provides adequate support. Place your knife blade ¹⁄₁₆ inch (2 mm) from the very tip of the nib and angled to roughly 55° to 60°. In one decisive movement, slice the end of the pen to finish the nib. Investigate this cut with a magnifying lens to check for rough texture or frayed wood fibers. If you find these, sharpen your knife and try again or smooth out the surface with fine sandpaper.

### Using the Bamboo Pen

Test drive your new pen to find out how it performs. Take some time to explore how it functions when you hold it in a variety of ways. Explore each of the available surfaces of the nib and the line properties that each part of the pen makes. Consider adding a reservoir and make any modifications you deem necessary to achieve the characteristics that you are looking for in your pen. And finally, make several reed pens of varying nib widths to play with in your creative explorations.

Bamboo pen in use with pine soot ink by the author.

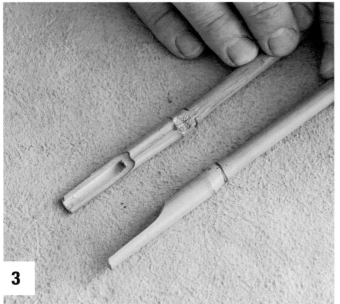

**3**

The first cut is made about 1¼ inches (3 cm) from the node end of the stalk. This oblique cut should be made to the depth of the center of the stalk.

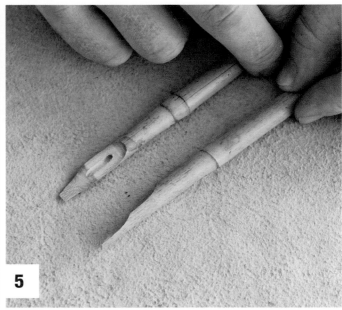

**5**

The second cut is made about ¾ inch (2 cm) from the nib end of the stalk, following the same oblique profile as the first cut.

**6**

Use a thin blade to make the center split about ¾ inch (2 cm) long. Place the blade edge in the center of the nib and gently rock it while applying small increments of downward/inward pressure until the split forms.

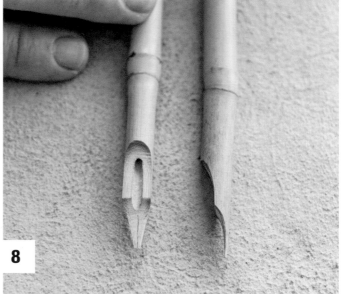

**8**

Make adjustments to either side of the split for symmetry. Then, with a sharp blade, make a 55° to 60° angle cut at the tip of the nib, from the outer wall side to achieve a finer pointed tip.

# Quill Pens

Quill pens have been used as writing implements for centuries, the number of which is not definitively known, although there are accounts of their use as early as the sixth century CE. Compared to reed pens, quill pens are flexible and responsive to light variations of touch and pressure. They have a gliding action when they are drawn across the surface of paper, creating lines that are smooth, fluid, and flexible. This sensitivity in responsiveness offers the ability to control the width of the line by varying downward pressure. The harder you push down, the wider the line gets. Artists took notice of these favorable properties in the twelfth century, when quills became preferred over reeds for drawing as well as writing. For most of the old masters, this was the case even as steel nib pens became widely available in the nineteenth century. These days quill pens get most of their action in the hands of illustrators and cartoonists.

Quill pens are fun to make and even more satisfying to use. They are made from the shafts of primary feathers (the outer-most wing feathers) from large birds, such as turkeys, geese, and swans.

Feather is a unique material to work with, and it requires you to spend some time getting acquainted with its distinctive characteristics. You will certainly learn of these properties as you gain experience working feathers as raw materials for your projects. Even so, I recommend setting aside the damaged, or less ideal, feathers to practice on. After shredding a few of these into smaller pieces, you'll be ready to make your first quill pen.

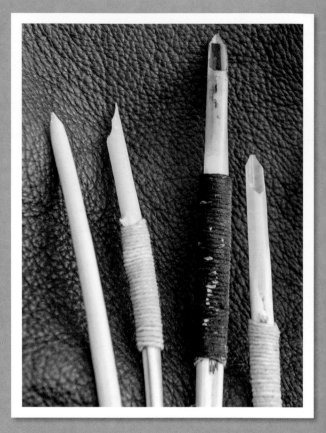

Turkey feather quill pens

## MATERIALS

Primary feather from turkey, goose, or other large bird (Note: Migratory birds are protected by law, and possession of feathers from some birds can result in a substantial fine, even if you find them on the ground. Check your local regulations.)

Fine sand

## TOOLS

Scissors
Sharp pocket knife or utility knife
Paper clip
Small metal bread loaf pan or tall metal can
Super fine sandpaper
Hot plate or stove-top burner

## 1: Get some feathers.

For quill pens, the primary wing feathers of turkeys, swans, geese, or other large birds is ideal. Keep in mind that the laws in your region may forbid the possession of feathers from specific species.

If you find smaller feathers along the way, don't discard them. You can use smaller feathers for other projects, so don't disregard them when they come into your life.

Introduce yourself to the farmers and hunters in your community. These folks may be good resources for those looking for feathers. Otherwise, you may be able to find some feathers online or at archery or craft supply stores.

## 2: Remove the barbs and the trim length.

The quill of the feather is referred to as the rachis, and it's flanked by the barbs, which is the technical term for the feathery part. Unless you want your pen to be adorned by the barbs, this is a good time to remove them from your future pen. Grip a section of feather barbs near the tip of the feather and pull down on it, peeling it toward the base, which is called the calamus. Repeat this on the opposite side of the feather. At this point the feather is stripped, leaving a naked quill. Use scissors or a knife to trim the quill to 8 to 9 inches (20.5 to 23 cm) long by removing a section from the thinner tip.

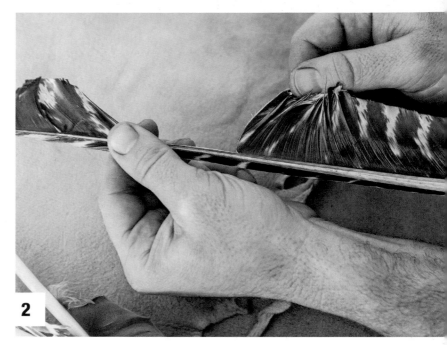

2

Remove the barbs of the feather by pinching them near the top of and tearing them off toward the base of the feather. These should come off in one strip, if not two or three. Repeat on the opposite side.

## 3: Cut the first bevel and remove the pith.

Cut the tip of the feather off at an angle approximately 1¼ inches (3 cm) long from the base of the quill, and only two-thirds through the cross-section. This cut is made on the underside of the feather's natural curve (see photo) and leaves the bottom half of the feather's cross-section intact. Once this cut is made, the hollow interior of the quill is exposed, revealing a pithy cellular material inside. Remove this pith with a slender tool such as a modified paper clip or even the tip of the quill that you removed earlier. If this material is not adequately removed or compressed deeper into the quill, it can interfere with ink flow in the finished pen.

3

Make a 1¼ inch (3 cm)-long slice at the base end of the quill at a diagonal angle starting 1¼ inches (3 cm) from the end. It should be clean and about ⅔ inch (1 cm) deep into the cross-section.

## 4: Heat treat the quill.

Fill a small bread pan with fine sand. Heat the sand up on a hot plate or stove-top burner until it is too hot to touch for more than a second or two. Turn off the heat and insert the prepared quill into the hot sand, nib end first, so it fills up with sand on the way in. Keep the quill in the hot sand for 5 minutes or so, or until it turns a slight yellowish color, compared to an untreated quill. At this stage the quill is rendered harder (more brittle), which makes the fine carving easier and creates a pen that requires less maintenance later when it comes to resharpening the tip. Next, using super fine sandpaper, your thumbnail, or other dull-edged tool, lightly scrape off any membranous tissue that remains on the outside of the quill.

Heat treat the quill by burying it in hot sand. Use a hot plate or stovetop burner to make the sand almost too hot to touch. Keep the quill in the sand for 5 minutes, or until it turns slightly yellow.

## 5: Square off the end.

Using a sharp pocket knife or utility knife, make a cut, perpendicular to the length of the quill, on the tip of the nib, leaving a flat end with sharp corner edges.

After the first cut has been made, and before moving on from heat treating, make sure the front and profile views of the quill look like this.

## 6: Split the nib.

Putting a shallow split (the slit) into the nib helps facilitate ink flow. Place a sharp, thin-bladed knife in the center of the squared tip, on the end of the quill. With your thumb on the back of the blade, carefully apply pressure while rocking the blade slightly back and forth, until the split begins to form. Gradually extend the split deeper until it is about 1 inch (2.5 cm) long.

Split the center of the tip with a sharp, thin-bladed knife. Make sure it does not extend beyond 1 inch (2.5 cm) in length. If it does, adjust the previous cuts to compensate.

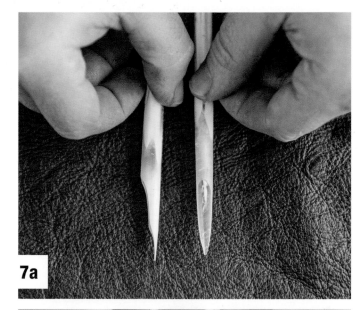

7a

Carve the tines of the nib to be symmetrical to one another, with the split as the centerline. This step is where you decide on how fine or bold you want your tip to be.

## 7: Narrow the tines.

Cut both sides of the quill ¾ inch (2 cm) from the tip to achieve the desired nib width. Strive for symmetry and keep the slit as the centerline. Make these cuts as clean and deliberate as possible, without leaving jagged or frayed edges.

## 8: Test your new pen.

Open a bottle of ink and try out your new pen. Modify the nib according to the line quality that you want from it.

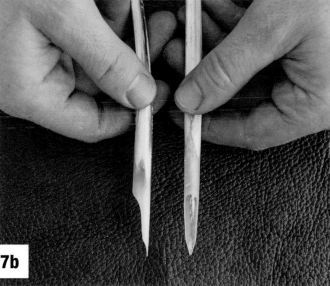

7b

Shape the angle of the nib from the top, or outer wall of the quill. Make this cut at a 60° to 65° angle, giving the very tip of the nib the sharpest point possible.

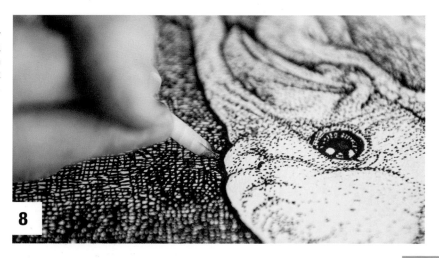

The author using a quill pen in a pen and ink drawing.

8

## Upgrades

There are many ways to accessorize your handmade reed and quill pens to fine-tune their performance and to customize them to your liking.

### INK RESERVOIR

The first thing I do to improve the function of a dip pen is to insert an ink reservoir. These are not essential, but they make it possible to work for longer periods of time without the need to recharge the pen in the inkwell. With extended pen use, the addition of a reservoir saves time and is well worth the extra effort required to make one.

Traditional reservoirs are made from thin, narrow sheets of brass, presumably because brass is corrosion resistant when used with iron/tannin inks. But I prefer to use birch bark, grape vine tendrils, or other rustic materials.

Fine-tuned pens with an assortment of embellishments. Left column: scroll pens, wide nibs, antler pen, ratcheting clamp pen with corked nib canister, bamboo nib jar. Right column: bamboo pen caps, scroll nibs for clamp pen, and fiber wrapped pen grips.

Here is how to do it: Find some birch bark that is thin in cross-section (comparable to cardstock paper), yet somewhat rigid and not floppy. Snip a tab slightly narrower in width than the tip of your nib, and up to 1 inch (2.5 cm) long. Next, roll one end of the reservoir tab around a small round object (such as the pen shaft) to train it to be curved. Do this on the other end of the tab as well, but this time in the opposite direction. The goal is to end up with an S-shaped tab. Maintain the S-shape and insert one bent end into the hollow portion of the nib (bend first, rather than end first), where it will wedge in securely. Position the reservoir so that the tip of the opposite end presses firmly against the nib, in contact with the slit and below the tip. This usually requires some fiddling with to get it just right. Be prepared to reorient, rebend, and snip excess length from the reservoir. The next thing to do is try it out and fine-tune it if necessary.

## VENT OR BREATHER HOLE

Modern dip pens have nibs with a little hole at the end of the slit, called the vent or breather hole. Vents allow ink to flow smoothly from the nib onto the paper. Another function of the vent hole is to protect the nib from splitting at the slit.

To make a vent in your reed or quill pens, you will need a paper clip and a candle. Light the candle and straighten a section of the paper clip, but leave one end as is for a handle. Heat the straight end of the paper clip in the candle flame until it begins to glow red. Next, place the hot tip ever-so-slightly beyond the end of the nib's slit, burning a shallow concavity. After a few seconds return the paper clip into the flame and repeat this process until it burns all the way through the wall of your nib. Now your pen has a vent hole.

## CORDAGE WRAP

Another simple thing to do to customize your reed or quill pen is to add a cordage wrap. This may aid in comfort and grip, and it will make your pen look really cool! Use thin diameter cordage so your pen doesn't get bulky. (See instructions in "The Knottless Lash" on page 25.)

Bend the reservoir into an S shape. Insert the reservoir so it settles in with one end in contact with the nib, just above the tip, so it does not get in the way in use. Modify the position as needed.

# Felt-Tipped/Clamp-Tipped Pen

My quest to devise a rustic felt-tipped pen led me to a fortuitous discovery. What I came up with is perhaps best described as a clamp-tip pen. This little device is a dip pen that uses a simple clamp mechanism mounted onto a pen shaft to allow the option for interchangeable tips. This is quite handy for experimenting with new nib or tip materials or variations of nibs that you already use, all without the need to make a whole new pen. In the world of dip pens, the clamp pen is an all-in-one multi-tool.

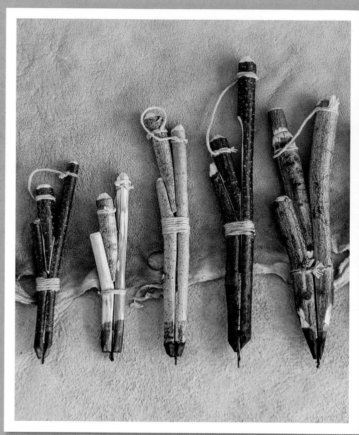

An arrangement of felt-tipped pens made from a variety of species

## MATERIALS

Three hardwood twigs at least ½ inch (1.3 cm) in diameter: one relatively straight one about 7 to 10 inches (18 to 25.5 cm) long, one with a bend in it about 2 to 4 inches (5 to 10 cm) long, and one 1 to 2 inches (2.5 to 5 cm) long

Strong cordage, such as 1.5 to 2 mm diameter hemp twine

Thin, rigid piece of birch bark

Felt or buckskin leather

## TOOLS

Sharp pocket knife
Heavy grit sandpaper
Scissors

## 1: Select the right sticks.

You will need strong sticks, such as those from hardwood trees, for this project, although the branches of so-called "softwood" trees will work just fine because they tend to be more dense than the wood that comes from the trunks of hardwood trees. You will need three sticks: one relatively straight one about 7 to 10 inches (18 to 25.5 cm) long for the pen shaft, one with a bend in it about 2 to 4 inches (5 to 10 cm) long for the clamp, and one straight piece 1 to 2 inches (2.5 to 5 cm) long for the wedge. (See photo above.)

## 2: Modify the sticks.

Let's deal with the pen shaft first. Using your pocket knife, carve a flat section (no more than one-third of the depth) along two-thirds of the length of the stick as shown. Next, carve a flat plane of the same depth along the convex edge of the bent stick that will be the clamp. Be sure the flat section you carve is on a consistent plane on either side of the bend. Finally, carve flat spots on both sides of one end of the wedge stick, making it somewhat triangular in shape from the side view, as shown.

1a

Look for a slightly bent stick to use for the clamp piece. Sighting down the stick reveals subtle contours.

1b

The raw materials of the felt-tipped pen, from left to right: the pen shaft, the clamp, and the wedge. Any strong wood of these dimensions will work well for making a pen.

2

When doing detailed carving, it's helpful to hold the knife with the thumbs on the back of the blade.

### 3: Carve the grooves.

The collar grooves are used as designated areas for lashing the pieces together. First, carve a collar into the clamp stick at the apex of the bend, being sure not to extend it onto the flat side. Next, line up the carved end of the pen shaft with the carved end of the bent clamp stick. Now mark the pen shaft stick where the bend of the clamp stick lines up when held together. This is where the next collar is to be carved. If these collars get too deep, (more than ⅛ inch [3 mm]) your pen will be weak in these areas. Two more collars are needed before we move on: one about ⅜ inch (1 cm) from the uncarved end of the pen shaft and the other ⅜ inch (1 cm) from the wider end of the wedge.

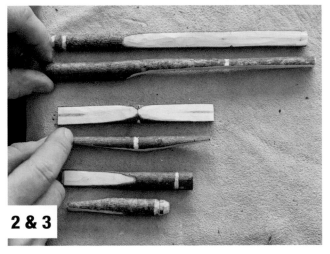

2 & 3

Carve a flat section along two-thirds of the pen shaft's length. Make this a consistent plane, without any twists. For the clamp: Carve two flat planes on the convex edge of the bent stick. For the wedge piece: Make wedgelike.

## 4: Lash the clamp to the pen shaft.

With some thin, yet strong cordage (such as hemp twine), secure the clamp stick to the pen shaft so their flat sides meet. This lash is the hinge for the clamp and should be as tight as possible, minimizing side-to-side action. Once tightly secured with a knotless lash, test the action of the hinge by rocking it back and forth.

**4a**

Begin the knotless lash by making a bend in the string while pressing it against the object to be lashed, in this case it is the pen shaft and clamp.

**4b**

Next, wrap over the bent segment of string with the longer end. Make this wrap as tight as possible and at least ¼ inch (6 mm) wide (5 to 10 times around). Be sure to leave the loop at the top exposed.

**4c**

To finish, insert the end of the string through the loop that you made in the beginning. Now pull on both ends of the string in opposing directions until the loop is tucked underneath the wrap. Trim the extra string.

**TIP:** Check to see how well these two flat planes meet. You want them to have as much surface area contact as possible. Take the time to adjust these planes in order to make it so.

## 5: Fit the wedge.

Now that you have a clamp built into the pen shaft, it is time to customize the wedge to make it functional. Squeeze the tip of the pen shaft together with the tip of the clamp, creating a wedge-shaped opening at the top of the clamp. Compare the dimensions of your wedge with that of the negative space at the top of the clamp. Using your knife, make any adjustments to the wedge that are needed to make the angles of these two shapes match more closely. This will make the wedge lock better into the clamp.

## 6: Lash the wedge to the top of the pen shaft.

The other two collars you made are for securing the wedge to the pen shaft as a permanent fixture so it does not get lost. Using more of the strong, fine cordage that you used earlier, tie a secure knot onto the end of the wedge. (A locking clove hitch is a good option.) Leave this cord roughly 4 to 6 inches (10 to 15 cm) long and tie the other end around the collar at the top of the pen shaft.

## 7: Shape the tips.

Use your knife to carve a taper to the ends of the pen shaft and clamp mouth. This makes your pen tip more refined and removes unnecessary mass. Be sure not to remove wood from the areas that you have previously carved flat. Instead, focus on the rounded, raw sides.

## 8: Cut the felt support.

The support for the felt is made from a thin, yet fairly rigid piece of birch bark. This is a resilient and flexible material that tends to hold its shape, yet is easily worked. To make the felt support, use a pair of scissors to cut a rectangle of birch bark as wide as you want the tip of your nib to be. Snip the length to be at least 1 inch (2.5 cm), long enough to fit into the clamp with at least ¼ inch (6 mm) protruding out at the tip of the pen.

## 9: Make the felt.

Using sharp scissors, snip a section of compressed wool felt (buckskin or suede works well too) to the same dimensions as you did for the felt support in step eight.

## 10: Insert the felt tip.

Lay the felt on top of the felt support with ⅟16 to ⅛ inch (2 to 3 mm) of felt protruding beyond the edge of the birch bark support. Now, with the wedge removed, open the clamp mouth at the tip of the pen. Insert the felt with the support into the clamp with the support on the far side (opposite the user), with at least ⅛ to ¼ inch (2 to 3 mm) sticking out beyond the tip of the pen. Insert the wedge tightly. Now it is ready to use!

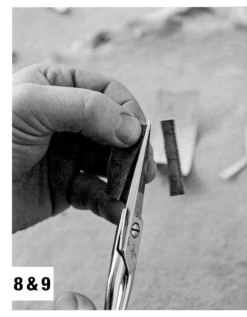

**8 & 9**

Snip a thin section of birch bark about 1 inch (2.5 cm) long. Now, snip a thin piece of felt or buckskin into the same dimensions. The width of these strips will determine the width of the line the pen will make.

**10**

Overlay the felt on top of the birch bark support tab with the felt extending slightly beyond the edge of the tab. Position these together in the clamp mouth and insert the wedge tightly, until the felt tip is snug.

The finished clamp pen with the wedge in place to hold the nib assembly.

## USING THE CLAMP PEN

A clamp pen is a versatile tool, and the tips used are limited only by your imagination. Below are some options for where to begin.

### Clamp pen with felt-tipped tab

For felt-tipped pens (markers), the nib is replaced by a piece of dense felt or buckskin leather, supported by a tab of birch bark. The felt is like a miniature sponge that absorbs and holds the ink, acting at once as the nib and the reservoir. Because you have the option for interchangeable felts, you can use narrow felts for fine lines and wider ones for more bold lines, all with the same pen.

When a felt tip is in place and when using undiluted ink, these pens work quite similar to markers, only you have to dip the pen to resupply the ink. This may seem cumbersome at first, but dipping the pen allows you to use the pen like a paintbrush as well. You can dilute the ink with water to achieve a wide range of tones between white and black. Just dab a spot of ink onto your palette (or paint dish) and add the amount of water that it takes to produce the tone that you are looking for.

Use as you do a dip pen and inkwell. When you are finished for the day, rinse the ink out of the felt with hot water so it does not stiffen up or, even worse, host a colony of mold.

Felt-tipped pen in use by the author.

### Clamp pen with birch bark–tab nib

If you want to get a different effect from the clamp pen, you can draw without the felt insert, using only the rigid birch bark tab instead. If the birch bark is firm enough to keep its form without bending too easily, it will provide you with a nib much like that of a calligraphy pen. These have the ability to draw lines as thin as the bark itself, and as thick as the width of the birch bark–tab nib. Whatever nib width you decide on, be sure to snip a ⅛-inch (3 mm) slit down the center to facilitate ink flow.

Inserting the birch bark–tab nib into the clamp pen: The nib consists of two pieces of birch bark, both of similar rigidity. The first piece, the nib tip, is the part that draws and is designed according to the line qualities you desire. The second piece, the sheath, is nearly twice as long as the first and is folded in half along its length. The tab nib is placed inside the sheath with the sheath folded over both sides of the nib, like a little sandwich. Make adjustments as necessary so the tab nib extends no more than ⅛ inch (3 mm) beyond the tab sheath. In addition to providing support for the tab nib, the sheath acts as an ink reservoir. Clamp this tab sandwich into the pen securely so it does not shift around while you are using it. Experiment with the thickness and width of the tab nib to get the effects and control you are looking for.

## Clamp pen as scroll pen

Scroll pens have nibs that are made with multiple tips. These create several lines at once as each tip of the nib draws across the surface of the paper. Again,

birch bark is an ideal material for crafting these nibs. For a scroll nib that makes two lines simultaneously, cut out a tab of birch bark twice as wide as what you would use for a single line tab (as in the nib pen option above). Next, snip a slit in the center of the tab, ¹⁄₁₆ inch (2 mm) long, dividing the tab into two squared halves. Snip away little triangles of birch bark from either side of one half until you have a triangular tip that is as sharp or blunt as you desire. Repeat this on the other side of the center slit to finish the scroll pen nib.

*Small Celebration of Paper Birch.* Nick Neddo 2013, birch bark–nib, felt-tipped pen, and pine soot ink drawing

## Modifying Clamp Pens

A useful way to modify the clamp pen is to hollow out the top of the pen shaft to add a felt/nib canister. Carve the top end of the pen shaft flat and mark the center of its cross-section with a concavity using a pocket knife. Choose an appropriate size drill bit and, placing it in the concavity, drill a 1¼-inch (3 cm) hole, keeping it centered. This takes practice and should be tried out first on less valuable material. Next, find a twig with the right diameter to use as a cork and whittle as needed. You can attach a leash to the cork just as with the wedge.

# 3

# INKS

We use ink all of the time. We write with it, print with it, draw with it, and see it all over the place. When you start making it, you will begin to see it from a whole new perspective. In this chapter, you will learn how to make inks from common materials and species on the landscape.

Inks are versatile as creative media because they can be used with a wide variety of artistic tools and they can be derived from many sources. Whether you prefer to use a pen, a brush, or even a stamp or other printing technique, you can use handmade ink.

Pen, brush, and ink self-portrait of author doodling with fresh inks. From left to right: coffee, beet, blackberry, black cherry, wild grape, buckthorn berry, elderberry, pine soot, poke berry, high bush cranberry, raspberry, blueberry, black walnut, and acorn

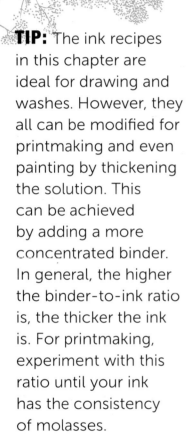

**TIP:** The ink recipes in this chapter are ideal for drawing and washes. However, they all can be modified for printmaking and even painting by thickening the solution. This can be achieved by adding a more concentrated binder. In general, the higher the binder-to-ink ratio is, the thicker the ink is. For printmaking, experiment with this ratio until your ink has the consistency of molasses.

# INK BASICS

Inks share some basic traits. All ink recipes have at least two ingredients in common: water and a pigment. The other ingredients have important roles, but may or may not be necessary, depending on the desired use. Inks are distinguished from paints by how dilute they are. Inks are more watered down than paints. Basically, if you dilute a paint enough to flow freely from a pen, you have a colored ink.

The basic ingredients of ink are few: a pigment, a binder, and sometimes a preservative and mordant, which is an agent that helps keep the pigment vibrant on the attached surface. Pigments are the primary and only necessary ingredient other than water. They come from many sources, including plants, animals, and even stones. We will get into pigments in a lot more detail in the "Pigments and Paints" chapter later on. (See page 103.)

Another important ingredient is the binder or vehicle. Binders keep particulates (pigments) more or less equally distributed in the solution (which is water). A short list of traditional binders is saliva, hide glue, gum arabic (resin from *Acacia* sp.), egg whites, oils, honey, and sometimes just water. (Water can be a solution and a binder. More on binders in the "Pigments and Paints" chapter on page 103.)

Depending on the source of pigment and your choice of binders, your ink may or may not need a preservative ingredient. Salt, essential oils, and aromatic herbs have been used in the past

**TIP:** In addition to adding ingredients to preserve your inks, the manner in which you store it affects its integrity. In general, all inks keep longer if kept in a cool, dark place. If you keep your inks out of direct light for the most part, you store them in clear glass jars. However, if your ink jars will be exposed to light in the storage area, use amber jars or other opaque containers, such as the options for inkwells presented in the next chapter. (See "Inkwells and Paint Dishes" on page 77.)

to help keep inks from getting funky and unpleasant to work with. Dehydrating the ink into a concentrated brick-like mass is another way to keep some inks from trying to decompose. These days commercial inks are preserved with fossil fuel-derived chemicals.

A mordant is used for inks that lose color and saturation over time , such as many berry inks. In this case, you can add vinegar to help keep the ink more stable over time. Alum (potassium aluminum sulfate), cream of tartar (potassium bitartrate), and ammonia are other mordant options, but they are less interesting to me due to the relative complexity of making them at home.

## INK HISTORY

Over time, people have experimented with a wide variety of plants, animals, and stones for the raw materials used to make ink. The juices from many fruits and vegetables offered the ink makers of the past a plethora of colors and properties to experiment with. The fluids found in squid and cuttlefish were the source of the original sepia. The decomposing caps of the inky cap (genus *Coprinus*) mushrooms have been used for centuries to make a functional ink. Tannic acid with iron was used for centuries as a common writing ink and gave us the basis for the preference of blue-black inks.

According to some historians, ink was first invented by the Ancient Egyptians. The oldest known book written in ink was found in Egypt, and it's about 4,600 years old. However, pottery shards decorated with ink have been found in Egypt that date even earlier. This ink was made of lampblack pigment and a base (binder) of gum arabic, a resin

*Oak and Acorn*. Nick Neddo 2013, pen, brush and acorn ink

from acacia trees. The Chinese have been making a similar ink since around the same time, 2,500 BCE. This ink was also made with a carbon black pigment but used a solution of hide glue as the binder. This ink is still being made today by Chinese ink makers in the same tradition and in a long lineage of fine craftsmanship.

Burned and crushed organic material has been the source of black pigment for inks as well as the soot gathered from burning resinous woods and animals fats. These black pigments are all part of a class of carbon particulates

called carbon blacks. Within this category are vegetable blacks, made from burnt vines and other vegetation; bone blacks, made from charred animal bones; and lampblacks, made from the soot collected from burning oil and fat in lamps.

Ultimately many of the juicy details of our ancient ancestors' relationship with inks are a mystery. Rediscovering some of these details through experimentation and creative playfulness is what this book is all about.

# HELPFUL TOOLS mentioned in this chapter

Paintbrush

Jars with lids

Hatchet

Knife

Stone mortar and pestle

Measuring cup

Slow cooker

Ceramic mortar and pestle

Small pot

Fine wire-mesh strainers

Canning funnel

Teaspoon

Metal bucket

Metal funnel

Saw

Tin snips or hack saw

# SIMPLE INKS

In my research and experiments in ink making, I have learned that ink can be made quite simply, with few ingredients and processes, or it can be complex in its manufacture, depending on its use and intended application. When it comes right down to it, inks can be made of just about anything that will stain your clothes. The difference between inks, dyes, and paints is somewhat arbitrary, or at best loosely defined.

The most substantial of differences is that paints tend to have a heavier or thicker binder than inks. When egg whites, oils, or other thick (or concentrated) binders are used, your ink becomes paint. Watercolors are an exception, because the binder used is water. Dyes are basically inks that are intended to color things that are exposed to more weathering, washing, drying, and general wear and tear. Because of this, dyes often have a strong mordant added to help keep the color from fading or washing out. Inks are traditionally used on supports (such as paper) that are fiercely protected from this kind of exposure.

Simple inks are classified (by me) to be inks that can be made quickly and that usually do not need the addition of binders or other ingredients. The following recipes are quick and easy, and they can offer endless hours of creative exploring and potential. We'll begin with some projects that are simple to make and are made from sources that are readily available.

Let's make some ink!

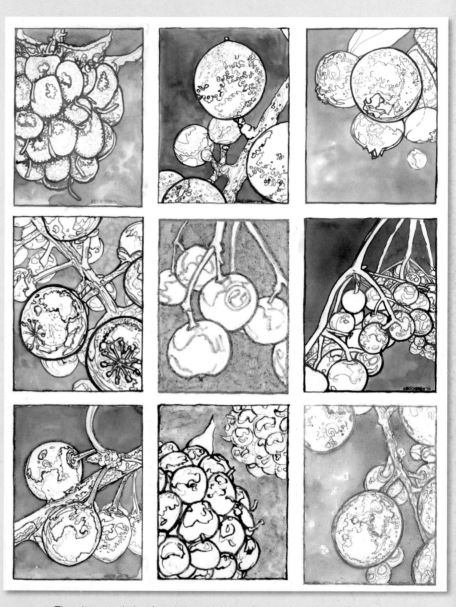

The characteristic of each berry ink is unique to its species. From left to right, top: raspberry, wild grape, and blueberry; middle: poke, high bush cranberry, and elderberry; bottom: buckthorn, blackberry, and black cherry.

# Coffee Ink

This sepia-tone ink is great for drawing with pens, doing washes with paintbrushes, writing, and even staining sheets of paper to give them a vintage look.

## MATERIALS

1 pot of coffee
  (4 to 8 coffee cups)
1 teaspoon vinegar
1 teaspoon salt

## TOOLS

Coffee pot
Filter
Mug
Small pot
Jar
Measuring cup
Teaspoon

Coffee stains do not have to be a bad thing. This piece is made by carefully controlling the coffee stain (ink!) with a pen and paintbrush on a sheet of paper.

## 1: Brew some coffee.

Get yourself some nice coffee and brew an extra strong pot of it. The stronger the coffee is, the darker the ink will be. When the pot of coffee finishes brewing, you have ink.

## 2: Pour some for yourself or a friend.

## 3: Reduce/concentrate the ink.

If you want your ink to be darker, you can transfer it to a small pot and reduce it on low heat until it has reached the saturation that you like.

## 4: Preserve it.

If you want your coffee ink to keep its value and have a longer shelf life, add 1 teaspoon of vinegar and 1 teaspoon of salt to each cup (235 ml) of coffee. The vinegar acts as a mordant, keeping the color from fading over time, and the salt is a preservative.

**TIP:** Set some ink aside to experiment with different binders. You may discover that the ink behaves entirely differently with binders than without.

# Beet Ink

Oddly enough, beets make a brownish color when they're simmered to make ink. To make a more brilliant, magenta-like ink from beets, juice them fresh, filter, add a binder and preservative, then bottle.

## MATERIALS

1 cup (225 g) chopped beets
1 teaspoon vinegar
1 teaspoon salt
1 pint (475 ml) water

## TOOLS

Sharp knife
Pot
Jar
Measuring cup
Teaspoon
Filter

When you cook a beet for an extended period of time, many of the nutrients go away, but the ink left behind is a joy to work with. Pen, brush, and beet ink.

## 1: Prepare and cook the beets.

Chop the beets fine, place them in a pot with water covering the beets, and boil them for 45 minutes or so. Once you have made sure that there is no nutritional value left in the poor things, boil them some more!

## 2: Reduce the solution.

Because the beets are already in the pot, you might as well reduce the solution down to the consistency and opaqueness that you want. Remember, the more the solution is reduced, the more opaque the ink will be. Boil it down to 1 cup (235 ml) or so. Then add the vinegar and salt to keep the ink around longer.

**TIP:** To get a darker ink, double the amount of chopped beets in this recipe.

# Berry Inks

Berry inks are fun and easy to make and use. Each berry species gives the ink maker a different color to work with that has its own unique behavior on paper and other kinds of supports. The recipe I have included here is featuring pokeweed berries (*Phytolacca americana*); however, any of the berries mentioned can be used in their place. Some of the berries mentioned here are toxic to eat, but they are safe for making inks. Regardless, keep the berries and ink safely out of reach of children. Each berry ink has a unique personality and spectrum of characteristics.

The amount of time it takes for the inks to dry is specific to each species of berry and which, if any, binders are in the mixture. For example, inks from black cherry (*Prunus serotina*) and common buckthorn (*Rhamnus cathartica*) that are prepared without a binder can take several weeks to dry on watercolor paper.

*A Cluster of Fresh Elderberries*. Nick Neddo 2013, pen, brush, and elderberry ink

## MATERIALS

1 cup (150 g) berries
1 teaspoon vinegar per ½ cup (120 ml) of berry extract (depending on the amount of juice extracted from berries)
1 teaspoon salt, per ½ cup (120 ml) of berry extract (depending on the amount of juice extracted from berries)

## TOOLS

Measuring cup
Teaspoon
Ceramic mortar and pestle
Fine mesh strainer
Canning funnel
Jar with tightly fitting lid
Stick or spoon
Small pot (optional)
Freezer (optional)

## 1: Get berries.

The more succulent and juicy they are, the better. The hardest part of making ink with berries is that you cannot eat them. To make matters worse, the qualities you look for in the most delicious berries are also what you want for making ink. In other words, the more vibrant, succulent, fresh, and local the berries are, the better the ink will be.

## 2: Freeze and thaw the berries.

This step is a great way to extract some of the berry juice without having to do much work. When the water in the cells of the berries freezes it expands as ice crystals. When this happens the cell walls are ruptured. When the berries thaw, some of the pigments inside are able to flow out of these broken cell walls and puddle up in the bottom of the container. This juice is of great quality for ink because it does not contain pulp or other solids that need to be strained out. Poor this berry juice into a jar and set it aside for later.

## 3: Crush berries.

Grind 1 cup (150 g) of berries with a ceramic mortar and pestle until you have a juicy pulp. Be thorough in order to get as much pigment out of them as possible, since eating them is not a good option anymore.

**TIP:** You can repeat the freeze/thaw cycle several times until the berries stop producing liquid. At that point it is time to move on to the next step.

Measure 1 cup (150 g) of juicy berries. They can be fresh or thawed from the freezer.

1

If you're using previously frozen berries, pour the juice through a fine strainer into a jar. This pigmented juice will be added to the ink that you will squeeze out of the berries in the next step.

2

Smash, smoosh, and grind the berries in a ceramic mortar and pestle in small batches. Be thorough to ensure maximum release of pigment.

3

## 4: Filter the berries.

Poor off the berry juice through a strainer and into a jar or other container that can fit a tight lid. Be patient; it can take a while for the liquid to percolate through the pulpy mass that accumulates in the strainer.

## 5: Add the vinegar.

To each ½ cup (120 ml) berry juice, add 1 teaspoon vinegar. This serves as both a mordant and a mild preservative.

## 6: Add the salt.

To each ½ cup (120 ml) berry juice, add 1 teaspoon salt. This serves as a preservative, helping to ensure that your new ink does not get funky—in a bad way.

## 7: Mix it all together.

If your ink is in a jar with a tight lid, you can shake it vigorously for a minute or so. Otherwise, stir it with a stick or spoon until the salt and vinegar have dissolved into the ink.

## 8: Store your ink.

Keep your berry inks out of direct light as much as possible. Plant derived inks are not lightfast. They tend to lose their brilliance and tone as they age in direct sunlight.

**4**

Use a canning funnel in conjunction with a fine-mesh strainer over a jar to filter the mashed berries from the precious berry juice. This may take a while, but be patient and let gravity do its job.

**TIP:** You may decide to filter you ink further to remove any remnant pulpy bits. One easy way to do this is to let the ink settle for several days. Then slowly pour the more pure liquid off from the top, through a fine filter, and into another jar. You will lose some quantity this way, but your ink will be more pure and less chunky.

Add 1 teaspoon vinegar to your berry ink as a preservative and mordant.

**5**

Add 1 teaspoon salt to your berry ink as an additional preservative. This step is worth it if you want your ink to last more than a few weeks.

**6**

The finished berry ink in a jar. This batch is made from poke berries (*Phytolacca americana*).

# LESS-SIMPLE INKS

The inks in this section are slightly more complex to make because they feature some special ingredients that require their own preparation: the iron/vinegar solution and the binder. The addition of an iron/vinegar solution to tannin-based inks creates an exciting chemical reaction. We will be using hide glue as a binder, although there are many other options. (See "Paint and Binders" in the "Pigments and Paints" chapter on page 103.)

*Small Celebration of Red Oak.* Nick Neddo 2013, pen, brush, and acorn ink. The bark, leaves, twigs, and acorns of the red oak tree, the source of the ink that was used to make the image.

# Acorn Ink

Inks from acorns fall into the iron tannate category of inks. These are inks made from extracting strong concentrations of tannic acid, which is a naturally occurring chemical in the plant world, and mixing it with iron in a vinegar solution.

You can use the acorns from oak trees (*Quercus* sp.) to make a fantastic ink. The color of acorn ink can be rich, warm, and dark. The magic ingredient in acorn ink is tannic acid, which can also be found in a wide variety of other plant and tree species. We will be using acorns because they are common, easy to collect, and cause no direct harm to the oak trees when harvested.

*Oak and Acorn.* Nick Neddo
2013, pen, brush, and acorn ink

## MATERIALS

16 cups (2.3 kg) acorns
1 teaspoon iron/vinegar
   solution (See "Make an
   Iron/Vinegar Solution"
   on page 64.)
1 teaspoon dissolved hide
   glue (see "Make Hide Glue"
   on page 65.)
Water

## TOOLS

Container to gather acorns
Stone mortar & pestle
Slow cooker
Strainer
Canning funnel
Measuring cup
Teaspoon
2 glass wide-mouth jars

**TIP:** Hide glue is gelatin. You can buy flavorless gelatin in grocery stores. Want to make your own? See the recipe for hide glue on page 65.

**TIP:** Other good sources of tannic acid are Eastern hemlock tree bark, alder bark, red dock root, sumac leaves, and even black tea!

## 1: Gather acorns.

Acorns can be found in the late summer among mature oak trees of any species. Some trees produce an abundance of acorns one year and quite less the next, depending on all sorts of variables that people try to understand. When you find a good spot to collect them, take a moment to enjoy where you are. Oak groves are some of the most majestic. The whole acorn is used, cap and all.

## 2: Crush the acorns.

Use the stone mortar and pestle to break open each of the acorns so they can be leached more effectively. You can skip this step if you don't mind boiling them for twice as long.

## 3: Extract the tannins.

Fill the slow cooker to the top with acorns and fill the cooker with water to within 1 inch (2.5 cm) of the rim. Put on the lid and be sure it is on snug. Turn on the heat. Simmer for at least 3 days, adding more water when it gets low. This acorn decoction will brew for several days to a week. Replenish the liquid level in the slow cooker before leaving your home for the day to prevent the ink from burning.

**TIP:** Set the slow cooker up outside somewhere. (I put mine on my porch, which is open to the outside.) This will prevent all of the moisture that evaporates off from causing damage (and drama) in your home.

## 4: Check the solution.

As the acorns simmer in the pot, the water level slowly diminishes. It's important to regularly refill the pot to the top as often as necessary so you do not burn your ink and damage your slow cooker or even worse, burn down your house!

## 5: Filter.

When you are satisfied that you have a super strong tannic acid solution, it's time to filter it. Place a canning funnel in a large mouth jar and put the strainer on top of it. Pour the solution through the strainer to remove the acorns and any other solid matter. (How can you tell? Taste it. It should be very astringent. If you don't want to taste it, you can just trust that several days of simmering is going to result in a strong tannin solution.)

*(continued on page 66)*

Gather acorns in whatever container you can get your hands on. This one is a bowl I made from a beaver log.

Smash open the acorns with a stone mortar and pestle. This will allow the tannins to extract faster, reducing the overall amount of time needed to simmer.

Fill the slow cooker with acorns and pour water in until it is 1 inch (2.5 cm) from the rim. Put the lid on and turn it on high. Simmer for at least three days, replacing the water when it gets low.

Allow the acorn solution to cool. Place a canning funnel in a large mouth jar and put the strainer on top of it. Carefully pour the solution through the strainer, separating the solids from the liquids.

## Make an Iron/ Vinegar Solution

## MATERIALS

Rusty metal bits and pieces

Vinegar

## TOOLS

Two jars

Stirring stick

To get started, gather some rusty metal objects. This little scavenger hunt might change the way you look at discarded metal. (Consider the process of making metal from dirt as a start.) I prefer nuts, bolts, screws, nails, and other rusty metal things that are comparable in size. Fill a jar or other container with these little treasures and designate a safe place to keep it.

When you're ready to make the solution, set

Gather rusty nuts, bolts and other small iron objects and put them in a jar with vinegar to let them oxidize. Alternate this liquid between two jars, pouring back and forth daily to let oxygen do its magic.

two jars next to each other, one empty and one containing the rusty iron treasures. Pour vinegar into this jar, covering all of the rusty metal pieces. The vinegar will accelerate the formation of new rust as it corrodes the metal back into its elemental ingredient: iron. At least once a day, pour the rusty vinegar into the second jar to allow the metal to be exposed to air. Rust forms much

faster in the presence of oxygen; the metal is literally oxidizing. The following day, pour the same vinegar back in with the metal. This is the process for creating and maintaining your very own rust nursery!

**TIP:** Be careful not to spill this anywhere near things that you do not want to rust.

## Make Hide Glue

People have been using animal skins, hooves, antlers, sinew, and connective tissue to make glue since the Stone Age. Our ancestors knew this substance well and simply called it "glue." (Only recently have we referred to it as hide glue.) It's derived from a mucus that coats and protects the collagenous tissue in the bodies of animals. The mucus is extracted from the animal tissue and refined into glue.

### MATERIALS

Animal skins, hooves, antlers, sinew, and other connective tissue
Water

### TOOLS

Knife
Large pot
Hot plate or stove
Strainer
Jar or other container

**1:** Prep the materials.

Gather the scraps of animal skins, antlers, hooves, sinew, and other miscellaneous connective tissue that you might have access to and cut them into small pieces. The smaller the pieces are, the more readily the mucus will extract from them.

**2:** Soak materials.

Put them into a pot and fill it with water.

**3:** Extract.

Using a hot plate (or your kitchen stove if you are single), bring your little witches brew to a low simmer using the lowest setting available. Critical detail alert: Don't let the temperature rise above 120°F (49°C) or your hide glue will be weak, literally. When it gets hotter than 120°F (49°C), the function of hide glue as a glue and as a binder is severely diminished. Let this slowly extract the mucus from these animal parts for several hours, adding in more water when necessary.

**4:** Strain.

Filter this collagen mucus extract (hide glue!) through a strainer into a different vessel. Place this in a cool, dry place where air is freely circulating to ensure that it does not spoil. As it cools and dries, it will set up and become quite gelatinous.

**5:** Cut it up.

Once the glue has set up like gelatin, slice it into tiny bits to allow it to dry further. Stir your gelatin bits periodically while they are drying to ensure they are not plotting any nasty secrets that have anything to do with decomposition. Do this until they are fully dry and hard like plastic.

**6:** Store.

Put the hide glue bits in a jar, bag, or other container where they can stay dry and await the occasion where they are needed for some really cool project.

**TIP:** You can buy unflavored, undyed gelatin from a grocery store and use it in place of homemade hide glue.

## 6: Add the iron/ vinegar solution.

At this point, you get to do some cool chemistry. Add 1 teaspoon iron/vinegar solution to each 1 cup (235 ml) acorn tannin solution. The color will quickly take on a darker hue. Stir or shake thoroughly for a minute or so.

## 7: Prepare the hide glue binder.

Add 1 teaspoon dried hide glue (see "Making Hide Glue" on page 65) to 1 cup (235 ml) water in a small pot. On the stove top, heat this up slowly (not getting hotter than 120°F [49°C]), stirring occasionally until the solids have dissolved. At this point, remove the pot from the heat and add 1 teaspoon hide glue solution to each 1 cup (235 ml) ink. Finally, stir or shake until it is mixed in.

**TIP:** If you don't have access to hide glue or commercial gelatin, you can use honey instead.

**6**

Add 1 teaspoon of iron/vinegar solution to your tannic acid extract. (See "Make an Iron/ Vinegar Solution" on page 64 for instructions on how to make one.) This is the catalyst for the chemical reaction that gives the ink its distinctive dark hue.

**7a**

Add 1 teaspoon hide glue (or gelatin) to 1 cup (235 ml) water and bring it up to 120°F (49°C). If it gets hotter than 120°F (49°C), the binder will be weak and will not perform well. Stir occasionally until dissolved.

**7b**

Add 1 teaspoon of hide glue solution to each 1 cup (235 ml) ink. This is a baseline measurement that you can modify to change the consistency of your ink to suit your needs. Stir or shake.

Finished acorn ink jarred and ready to use

# Black Walnut Ink

This exquisite brown ink resembles sepia (made from cuttlefish), and it can be used concentrated for dark, bold lines, or diluted for lighter, more tan tones.

## MATERIALS

16 cups (1 kg) black walnut hulls
1 teaspoon iron/vinegar solution (optional)
Water

## TOOLS

Stone mortar and pestle
Slow cooker
Strainer
Canning funnel
Measuring cup
Teaspoon
Jars

*Looking up a Walnut Tree*. Nick Neddo 2013, pen and ink. Both the pen and the ink were made from the black walnut tree.

## 1: Find a tree.

Look for a mature, nut-bearing black walnut (*Juglans nigra*) tree.

## 2: Gather walnuts.

It's ideal to do this in late summer when they are fresh.

Gather black walnuts in late summer when they are fresh.

### 3: Separate the hulls from the nutshells.

The hull of the nut is where all the ink is. You need to remove the hulls from the nuts by smashing them with the stone mortar and pestle. Once the hulls have been broken, you can separate them from the woody nut shell.

**TIP:** Black walnut hulls can stain your hands and clothes if you aren't careful. After all, they are full of ink. Wear gloves and perhaps clothes that you don't mind if they take on some more character.

3

Smash the hulls open in the stone mortar and pestle to separate the fleshy hulls from the walnut and shell inside. This step is much easier when the walnuts are fresh and still moist.

Dry the extra walnut hulls in the sun or somewhere dry where air is circulating. These can be used for inks and dyes later.

## 4: Simmer.

Fill the slow cooker with the hulls, and then cover them with water. Simmer the hulls for several hours to several days, depending on how much ink you want to extract from them. Refill the slow cooker with water as needed, being careful not to let it all boil away and burn your ink or your house.

## 5: Reduce the walnut extract.

Continue simmering the hulls until you have about 1 pint (475 ml) of liquid left in the slow cooker.

## 6: Strain the solids.

Let the ink cool. Then pour it through a filter to separate out the hull solids. You can repeat this process several times with finer filters if you wish.

## 7: Add hide glue binder.

Add 1 teaspoon hide glue solution to each 1 cup (235 ml) extract and mix well.

Simmer the black walnuts in a slow cooker (see "Acorn Ink" on page 61) for at least three days, making sure that you replenish the water level as it gets low.

**4**

Filter the ink through a strainer, separating the solid walnut hulls away from the ink. Add 1 teaspoon of hide glue solution to each 1 cup of extract (see "Acorn Ink" on page 61) and mix well.

**6**

**8** Darken ink (optional).

If you want a darker ink from your walnuts, you can add 1 teaspoon iron/vinegar solution to each 1 cup (235 ml) ink. I like to set some aside first to keep unadulterated and use the iron with the rest.

**9:** Bottle your ink.

Any jar or bottle with an airtight seal will work well for long-term ink storage. Choose amber or other dark colored glass to keep light from coming in. If clear glass is used, keep the ink stored in a dark place when not in use.

Finished black walnut ink in a jar and ready to use.

**9**

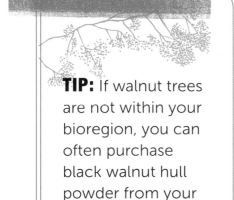

**TIP:** If walnut trees are not within your bioregion, you can often purchase black walnut hull powder from your local herbalist. This form of walnut hull is ideal to work with.

*Small Celebration of Black Walnut.* Nick Neddo 2013, pen, brush and black walnut ink. The bark, leaves, twigs, and nuts of the black walnut, the source of the ink that was used to make the image.

# Soot Ink

Soot is in a class of super fine carbon particles that are created as a result of inefficient combustion—burning resinous material in at low oxygen environment. Soot is the source of pigment in black inks known as India ink and Chinese ink. One type of soot is called lampblack because it is created as a result of burning an oil lamp under a plate or other surface in which the soot can accumulate. The soot from burning resinous pine wood is called pineblack.

This is the soot that we'll use in the following recipe, although any fine soot will work well.

This ink recipe requires a prerequisite step. (See "Prerequisite: Make a Soot Collector" on page 73.)

*Greeting a Friend.* Nick Neddo 2013, pine soot ink wash

## MATERIALS

Pineblack (soot)
Hide glue
Water
Rosemary essential oil
   (optional)

## TOOLS

Soot collector (See
   "Prerequesite: Make a Soot
   Collector" on page 73.)
Hatchet or saw
Dust mask
Small paintbrush, firm
   feather, or other tool
Jar with sealable lid
Porcelain or glass mortar
   and pestle
Fire

## 1: Gather and process soot materials.

Once you have devised a soot collector, it's time to find some resinous material to burn in it.

Any natural material that creates copious amounts of thick, dark smoke when burned is worth consideration as a source of soot. Resinous woods from trees in the pine family (*Pinus* spp.) are great for making soot. The knots (or branch bases) in the decayed stumps of dead trees are often saturated with pitch (called lightwood), which produces copious amounts of soot compared to other woods.

To access the lightwood, you will have to excavate the soft punky wood away from the rot-resistant lightwood knots. Once you have exposed these fins of dense wood, you may need a hatchet or saw to break off chunks of this precious stuff. Scrape or carve away any soft, rotten wood from around the lightwood and chop, saw, or break the pieces into smaller chunks (no larger than your hand). Next, place them in the sunlight to dry.

Find an old pine stump that is soft and well on its way to being soil again. Inside, you will find the dense branch knots, preserved and saturated with pitch. Use a hatchet or saw to get these out.

## Prerequisite: Make a Soot Collector

Soot collectors can be simple or elaborate. You can make a soot collector quite easily. There are many possibilities and variations to the design, but the principals are the same. You will need a metal cylinder for the shell and a metal funnel for the lid. The funnel lid needs to be at least slightly larger in circumference than the shell is, or it will not fit on top. The goal is to make a container for a fire, with a surface directly above the smoke and flames. This is one instance when you intentionally want a smoky fire. The soot (from the smoke) collects on the walls and ceiling of your soot collector as you burn a smoky fire.

MATERIALS

Old metal bucket
Large metal funnel that
    can fit over the bucket

TOOLS

Hacksaw or tin snips
Metal file, optional

**1:** Remove the bottom of the bucket.

Use the hacksaw or tin snips to cut the bottom off of the metal bucket. Be careful! You may want to file down or remove any sharp edges that remain to avoid getting cut. Use a metal file if this is an issue.

**2:** Add the metal funnel to the bucket shell.

You have a soot collector! Now you are ready to make some really amazing ink. It's my personal favorite for sure.

Place a metal funnel on top of the metal cylinder to give all that sooty smoke somewhere to condense. Now stand back and watch this short-lived but exuberant smoke show.

**TIP:** The fresh cones from pine trees are often covered with pitch, and you can use them in addition to or in place of the old stump knots. These are often much easier to collect, and they only require drying before use. The papery bark of birch (*Betula* sp.) trees is full of a waxy resinous substance that creates a deep, dark smoke when burned. This makes a fine soot I call birchblack. In addition, the fats from animals make a good pigment soot when burned in an oil lamp. Be creative and have fun.

*Small Celebration of White Pine*. Nick Neddo 2013, pen, brush, and pine soot ink. The bark, leaves, twigs, and cones of the Eastern white pine, the source of the ink that was used to make the image.

## 2: Make some soot.

Light a fire in your soot collector using a resinous material (see above). After the fire has gone out and the soot collector has cooled off, put on your dust mask. Now you can gather the lampblack with a paintbrush, firm feather, or other such tool. This is the part of the process where the funnel top of the soot collector comes in handy. Invert the funnel into a jar and carefully remove the fine layer of soot that has accumulated. As you do this the soot will fall down the funnel and into your jar.

**2a**

Cover the fire with the metal cylinder you made from an old bucket.

**2b**

When the soot collector has cooled off, put on your dust mask and then remove the funnel from the cylinder and invert it into a small jar. Use a paintbrush to remove the soot from the inside of the funnel and guide it into the jar.

## 3: Mix and grind with hide glue solution.

In your smooth mortar and pestle, add a small amount (perhaps a teaspoon) of soot powder. Next, add an equal amount of hide glue solution (dissolve 1 teaspoon to each 1 cup [235 ml] of water, and lightly heat in a pot until the hide glue is dissolved in the water). Carefully grind the mixture, adding in more soot as needed, until all of the particulates are saturated and there are no longer any dried clumps. Soot is hydrophobic, which means it repels water, and it mixes evenly in the glue solution only after a significant effort of manual grinding. This process can take several hours to days and is best done intermittently. I like to use this process as a sort of meditation, but other times I'll put on an audio book while grinding carbon black inks. The goal is to get all of the particulates to sink and to have even consistency in the texture of your ink.

## 4: Add preservatives.

When you are satisfied with the consistency of your ink, add a couple drops of rosemary essential oil to keep it from getting stinky as it ages. If at any time you begin to dislike the smell, add more essential oils.

## 5: Bottle your soot ink.

Keep a tight lid on it so you do not accidentally spill any of this extremely precious liquid.

Put ½ cup (120 ml) of soot powder into a ceramic mortar and pestle, being careful not to spill any of this precious material.

**3a**

Add 1 teaspoon hide glue solution (see "Acorn Ink" on page 61) to ½ cup (120 ml) water and slowly pour it into the mortar and pestle with the pine soot.

**3b**

**3c**

Grind the soot in a ceramic mortar and pestle to blend the carbon with the water and hide glue. Put your meditation hat on because this can take between 3 and 30 hours, depending on the purity desired.

4

# INKWELLS AND PAINT DISHES

You can make inkwells and paint dishes from a wide variety of materials. Pictured: bone, clay, basswood, bamboo, stone, and shell

Now that you've made some ink, you may want to upgrade the vessel that you keep it in. Little jars with tightly fitting lids are perfectly suited to the job, but if you want your inkwell to be a work of art, consider making your own.

Paint dishes are also great projects to make out of clay. Then, of course, there is the whole world of sculptural ceramics. Be creative and have fun.

# Clay Vessels

Clay (and the resulting ceramic) as a material for artist tool making is unrivaled in its utility and creative potential. Pottery projects can be quite therapeutic and intuitive as well as challenging and intellectual, all depending on your approach. Many high quality commercial clays are available, fully tempered and ready to use. The following information will be useful for people who want to know how to process and use the local wild clays in their bioregions.

A word of caution: The pursuit of the secrets of wild clay often is a process of fierce patience and non-attachment. Many variables can result in your precious clay pots cracking, fragmenting, crumbling, and even exploding. Try to put yourself in the shoes (or moccasins) of your ancestors when they were discovering the magic of transforming mud into ceramic. Use your failures as learning opportunities, or just buy commercial clay and fire it in a kiln.

Raw, unprocessed clay in its natural form. Clay is one of the most abundant materials on Earth, and it comes in a variety of colors with a range of purity.

## MATERIALS

Clay
1 cup (235 ml) water
Temper (sand works well)
Firewood

## TOOLS

Thick leather gloves
Shovel to dig fire pit
Metal rock rake
Hose or other source of
    emergency water
Ground mat
Mussel shell (optional)
Wooden paddle (optional)
Burnishing stone
    (optional)

Tools used in this chapter: fire, hacksaw, table vise, knife, metal rake, and thick leather gloves

## 1: Get some clay.

You can find clay near road cuts, rivers, and streams throughout the world. You want to find a clean and pure source, where there is no sand or organic material mixed within it. The presence of these impurities in the clay body can cause problems when hand-building and firing it.

**TIP:** Good quality clays can be made into a coil and tied into a loose overhand knot (like a pretzel) without breaking or excessive cracking. This will be the first test of the clay in question.

## 2: Wedge and knead the clay.

The process of making clay workable will depend on if your local clay is dry and hard or damp and not so hard. To make raw clay workable, it must first be pounded, kneaded, and worked until it loosens up and resembles the workability of commercial clay. You may need to add small amounts of water to the clay body as you do this.

**TIPS:**

Be sure not to add too much water or else you will have to wait several hours or days before the clay is ready to work again.

If your local wild clay is dry, as it often is in arid regions, pound it into a powder with a hammer, stone, or a stone mortar and pestle. Next, sift it through a mesh strainer to remove possible impurities. You can also purify the clay with the same method as purifying pigments. (See "Mineral Pigments and Paints" on page 108.) Rehydrate the powdered clay by adding small amounts of water and then kneading it. Continue the wedging process until the texture is silky and uniform.

Test the workability of found clay by making a coil and then tying it into a simple overhand knot. If cracks and fissures appear, the clay may be too dry, or it may contain too many impurities to work with.

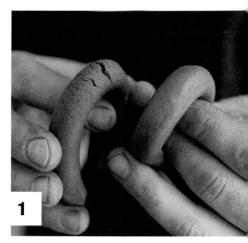

Wedge raw clay in small batches by kneading and throwing it onto a hard, flat surface. Work the clay in this manner until it becomes consistently elastic. Be thorough and try not to trap air bubbles within the clay body.

## 3: Temper the clay.

Once the clay is workable, you need to temper it. (If you are using pretempered clay, this isn't necessary.) When clay dries, it shrinks, and without temper in your clay body, it will crack as it does so. Temper allows the clay to shrink without cracking. Sand makes a reliable temper, but finely crushed pottery shards (called grog) are used in many commercial clays. Depending on the qualities of the clay you find, the ideal ratio of temper-to-clay will range in the 0 to 30 percent range. It's difficult to predict what percentage of temper you will need in the clay body, and for this reason I recommend making some test tiles. (See "Make Test Tiles with Your Found Clay" below.)

When adding temper to the clay, be sure to knead it all together thoroughly and gradually. The more consistent the distribution of temper is in the clay, the more likely it is that your vessel will survive the coming ordeal, which is the firing process.

**3**

For 30 percent temper, make two balls of clay of equal size and a pile of temper of a comparable mass. For 25 percent, make three balls; 20 percent, four balls and so on. Wedge thoroughly, mixing the temper into the clay evenly.

## Make Test Tiles with Your Found Clay

Your tiles should be consistent in their dimensions and clearly marked with their temper percentages. I make mine about 1 centimeter thick, by 4 centimeters long, by 4 centimeters wide, and I scratch the temper percentage number on the top surface. You'll see some results even before you fire the tiles. If cracks and fissures develop as the clay dries, there is not enough temper. After these tiles are fired, you'll have some great feedback as to the best temper ratio for the specific clay you are working with.

It's a great idea to make at least two of each tile to gather better data from your research.

Another option is to make small test pots in place of test tiles. These act as test tiles in the way that they are small and made relatively quickly, but with good craftsmanship. If these "test pots" fire successfully, they have a much more useful life as a vessel rather than a ceramic disc.

## 4: Hand build.

It's time to make a pinch pot. Start with a ball of clay the size of an apple. Wrap the rest of your clay in a plastic bag or large leaf to prevent it from drying out too much while you're working. Wedge your ball of clay a bit more, getting out all of the air bubbles, and form the ball again. Next, push your thumb into it while turning it in your hands. As you rotate the clay ball, insert your thumb gradually deeper until you have made it about three-quarters of the way through. Keep your thumb moist while you work. From here, you can make space for your other thumb and work on thinning the walls of your vessel as the internal space increases. There are infinite design options and creative embellishments, but the important thing is that they are functional. Be sure to have relative consistency in the thickness of the walls and base and fix small cracks as they form.

## 5: Dry the vessel.

Allow your clay creations to dry slowly for the next few weeks. Place them on a board away from direct sunlight and rotate them every so often. This will allow for even drying throughout the piece.

Make small vessels to whatever form you're inspired to create. For inkwells, be sure a cork or lid of some sort can be used. Finished pots need to be set aside in a protected place for drying.

**TIP:** If the clay is cracking while you work, it may be too dry. Dip your fingers in water and repair cracks by smoothing them back and forth with your finger.

**TIP:** You can use the oven to dry your pots. When the pots appear to be dry (after a week or two, depending on the humidity in the air), put them in the oven on its lowest heat setting (150 to 200°F [65 to 93°C]). Each half hour or so, increase the temperature by 25 to 50°F (20 to 30°C) until it is maxed out at the highest temperature (broil) for 1 hour. Turn the oven off and proceed to prefiring.

## 6: Prep the fire pit.

For your pottery "kiln" dig an oval-shaped fire pit, oriented parallel to the prevailing wind patterns in the area. Make it about 4 feet (1.2 m) long, 3 feet (0.9 m) wide, and about 6 inches (15 cm) deep on the downwind side of the oval. Use the dirt that you excavate to build a berm on the downwind side, in effect making it twice as deep on that side. The opposite side of the oval should gradually slope up to match the surrounding ground level. This shape acts as a wind tunnel when a fire is present and is essentially what was used by people in the early centuries of the ceramic industry.

Prepare the firing area by organizing firewood into size-graded piles and gathering the tools: rock rake, thick leather gloves, hose or other source of emergency water, and ground mat. Light the fire to dry the area before bringing the pots outside.

## 7: Prefire.

The following information is relevant to open-pit firings in a large campfire. Once the vessels are thoroughly dry (usually after one to three weeks of air drying), you are ready to prefire them. The success or failure of your pottery is largely dependent on how this step goes. The length of time it takes for all of the water to leave your clay pots depends on how large your vessels are and how thick their walls are. Be patient! Give yourself the time to do this right. Let your pots prefire for as long as you can, for at least a couple of hours to give them the best chance of surviving the thermal shock of turning into ceramic.

When the ground surrounding the fire pit is dry, arrange the pots at a safe distance from extreme heat (no closer than you could comfortably keep your hand). Rotate the pots every which way at this comfortably warm distance, making sure all surfaces are exposed to the radiant heat.

First, light a fire in the location where you will be firing you pots about an hour or so before you place the pots on the ground in the periphery of the fire. This ensures that the ground where your pots will be placed is dry. Place the vessels on a mat if you're concerned about them taking on moisture from the ground. At this early stage of prefiring, you don't want to have the pots any closer to the fire than where you could keep your hand without burning it. Keep the pots at this distance from the fire, rotating them methodically (side to side, front to back) every few minutes to ensure even heat exposure. After 10 to 30 minutes, move them a little closer to the fire and repeat the rotation process. It won't be long before the pots become too hot to handle, and you'll need to wear thick leather gloves to continue rotating and moving the pots. Prefiring is complete when the pots are nearly touching the fire, being rotated regularly, and have been that way for at least ½ hour. At this point, you're ready to fire the pots.

7b

After 30 minutes or so, move the pots a few inches closer to the fire, maintaining it at a consistent heat output. Continue to methodically rotate and turn each pot to ensure even heat exposure on all surfaces.

7c

Continue this process of systematically rotating the pots and moving them closer over the next 2 to 3 hours. They will become too hot to touch without gloves. Even so, continue prefiring until they are very close to, but not in, the fire.

7d

When prefiring is complete, remove the fire from the pit by raking the coals to the outside. Clear away any combustible material from the floor of the fire pit so it does not reignite before you are ready.

Inkwells and Paint Dishes

## 8: Fire!

The transition between prefiring and firing is time sensitive, and it has to be approached with a clear plan for your sequence of operations. Before you make a move to fire your pots, be sure to have a large bed of coals in your fire pit to work with. Also, have all of your firewood readily available and organized in piles of relative size (kindling, finger-diameter, wrist-diameter, etc.). Ready? Okay, still wearing your gloves, rake the coals aside to the far edges of the fire pit to make room for the pottery platform.

Make your pottery platform by placing 2 to 3 foot (0.6 to 0.9 m)-long, 1 inch (2.5 cm)-diameter dry sticks parallel to one another in the center of your oval-shaped fire pit. Next, place similar sticks of half that length on top of the long ones in a perpendicular orientation, making a sort of grid structure. Make it tight enough to prevent your little pots from falling through. This is the platform for your pots. Carefully place your pots onto the platform, keeping them close to each other, but not touching.

Build the fire structure with all the wood you plan on using so you do not have to tend it during firing. Choose some of your finest kindling twigs to place in between, around, and on top of your pots. Then, use roughly half of your next smallest firewood to place around and on top of the kindling. Place the firewood onto this structure in gradients of increasing size until you have created somewhat of a

**8a**

Build a lattice-work platform for your pots, using long, straight firewood of a consistent diameter. Arrange them lengthwise on the floor of the pit. Make another row on top by placing shorter pieces of similar diameter perpendicular to the first.

**8b**

Wear gloves to arrange the pots close together in the center of the platform. They can be touching or separated by small pieces of dry firewood. Be efficient, yet intentional.

**8c**

Use fine kindling to cover the pots with a dense cushion of highly combustible material.

**8d**

Next, layer pencil-diameter firewood on top of the kindling cushion. Build the structure in a conical shape, adding increments of larger wood in each layer.

monstrosity of a tipi fire, with the largest of your firewood on the outside. Take the kindling that you set aside and place it at the base of the structure, as well as in between each of the largest pieces of firewood. Be sure to shove the kindling in toward the center to ensure more rapid ignition. Finally, rake the coals back in around the fire structure amidst the kindling until it takes flame. Now you can give your precious pots one more wish of good luck as you step back and watch the conflagration.

The clay pots transform into ceramic only after they glow red hot, or incandesce. This metamorphosis is referred to as ceramic change, and the temperature that causes it varies according to the clay. Because ceramic change is a temperature-based phenomenon, the amount of time you fire your pots is less important than how hot you can make your fire. The design of the fire pit can greatly affect how hot your fire is.

Another factor to consider is what species of wood you are burning. In general, hardwoods burn longer than softwoods, put off less light, and have more persistent coals. Softwoods burn faster and hotter, and they give more light. For this reason, I tend to use softwoods more often than hardwoods.

**8e**

Build the oversized tipi fire with symmetrical architecture, leaving ignition doorways around the base. After all of the large wood is in place, stuff each doorway with the remaining fine twigs and rake the coals back to reignite the structure.

**8f**

As the pottery fire burns, don't add wood to it. The pots are still quite vulnerable, and they can be broken as the fire structure shifts. Be ready to control the conflagration if it spreads beyond the pit.

**TIP:** Be prepared to control the fire if it gets out of hand. I like to have a hose or at least buckets of water nearby, just in case.

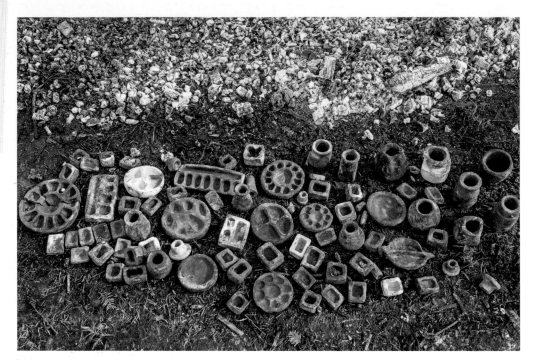

Small pots, freshly initiated into the ceramics club after firing. Notice the fire-clouding, the distinctive patterning unique to pottery directly fired by wood. Do not be too anxious to handle newly fired pots, unless you are willing to get burned.

## Trouble Shooting:
## Failed Pot During Firing

When pots fail, it's usually the result of one or more of three common variables.

- The quality of the clay being used: If the clay has too many impurities, it may fail in the firing or become crumbly after firing.

- Poor building technique: Another cause of vessel failure is due to poor building technique by the potter. If the wall thickness is widely variable or if fissures are left unattended, the vessel may break during firing.

- Hasty prefiring: The most common cause of failed pottery vessels in my experience is due to hasty prefiring, or incomplete prefiring. Water that is still bound within the clay will surely cause the pot to fail in the firing. Be thorough and patient in the prefiring process! "Haste makes waste."

—A smart person said this.

# Bone Inkwells

Bone is a wonderful material to work with for making and using all manner of tools and small vessels. It is a durable, shapeable material that's essentially a composition of minerals collected by plants and concentrated by animals.

## MATERIALS

Large bone
Piece of soft- to medium-
    density wood
Piece of charcoal
Waterproof wood glue or
    beeswax

## TOOLS

Hacksaw
Light mallet
Sharp knife
Sandpaper
Wood or folding saw

An assortment of bone inkwells and processed bones. The leg bones of cows, deer, moose, and other large mammals are ideal for making inkwells. You can find them at pet stores sold as chew toys for dogs.

## 1: Get a bone.

For making inkwells from bone, you will need to find a large hollow bone, usually a leg bone, from a large mammal such as a cow, moose, or buffalo. You can find them on the ground after minutes to years of wandering, depending on where you wander. If you lack the patience to search out the skeletons of animals from the wild landscape, you can buy good quality bones at a pet store. The large cow bones they sell for dogs are ideal. Sometimes you can buy one that is cut to about 2 inches (5 cm) in length already, other times you will have to buy a longer bone and cut it into shorter lengths to work with.

## 2: Hacksaw the bone to length.

If you need to reduce the length of the bone that you're working with, a hacksaw is the tool for the job. If you have a table vise, you can hold the bone in that while you saw. Otherwise, you can set yourself up on a bench or other elevated surface with your foot holding the bone as you saw. (See photo.)

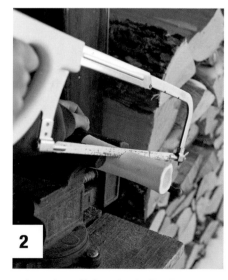

**2**

A section of bone that is 2 to 3 inches (5 to 7.5 cm) in length is suitable. If you have a 5- to 6-inch (12.5 to 15 cm) bone, cut it in half with a hacksaw. Use a table vise or other clamp mechanism to hold it in place as you saw.

## 3: Make the base plate cork.

The bone is hollow all the way through and that includes the base. We are going to have to make a base plate from wood, preferably from a tree species that is easy to work with, such as cedar, willow, linden, aspen, birch, pine, or any other soft to medium density wood. This base plate cork has to fit as snug as possible so it doesn't leak. This brings us to the art of what I call "snug-cork-fittery," which is relevant to the base plate as well as the top cork.

This art has its own set of steps.

First, get some suitable wood. This should be dry because wet (green) wood will change dimensions as it dries, causing your cork or base plate to not fit the way it did when you made it. Cut the wood to at least 12 inches (30.5 cm) in length so you have something to hold on to while you whittle.

Next, using your knife, whittle one end of the stick down to just slightly larger in diameter than that of the inner dimensions of the bone's opening. Do your best to make the cross-section of the stick match the shape of the bone's negative space (the socket). Because the bone socket is a unique shape, you will need to mark the base plate stick somehow in relation to the bone so you are inserting it into the socket in the same orientation each time you test its fit. Once the end of the stick begins to fit into the socket, it's time to scribe.

Mark the inside of the bone with a piece of charcoal where you want the base plate to fit. Now insert the stick into the scribed opening as far as it will allow you to at this point, and then remove it. The end of the stick will be marked with charcoal, revealing where to focus your carving. Whittle away all scribed areas and refit the stick once again. Each time you do this, you will be able to fit the cork into the socket with more surface area contact than the previous time. The goal here is to repeat this process (reapplying charcoal as needed) until the stick comes out fully covered with charcoal, indicating that the cork is in maximum contact with the socket.

Finally, you can apply waterproof wood glue to the outside of the cork stick and fit it into the bone socket in its proper orientation. Using a mallet or larger stick, lightly tap it in a bit deeper than what you can manage by hand to ensure a permanent fit. The last steps for the base plate are to saw off the extra length where the stick protrudes from the bone with a wood or folding saw, and then sand the bottom smooth if you want to.

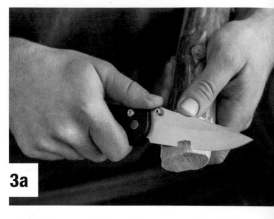

Cut the base cork branch long enough to hold on to while you whittle it to shape. Make the end of the stick resemble the shape of the bone socket, and small enough in diameter to begin to fit into it.

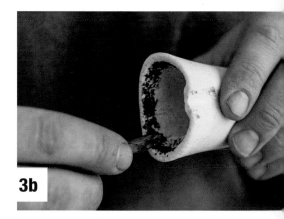

Use a piece of charcoal to mark (scribe) the inside diameter of the bone socket at its opening.

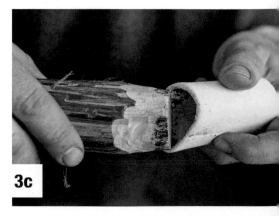

As the base-cork takes shape, insert it into the scrib[e] bone socket as best as you can.

**3d**

Remove the base-cork piece from the socket to reveal the places that are marked by charcoal. Carve these scribed areas off and test the fit again, keeping a consistent orientation. Repeat this process with patience until the base-cork fits perfectly.

**3f**

Use a folding saw to trim off the excess wood from the base-cork. Make the cut close to the base of the bone so you don't have a lot of mass to sand off later.

When you have a snug fit, pound the base-cork into the bone socket a little deeper with a light mallet. Use waterproof wood glue to seal the seam if you are not confident with the integrity of this union.

**3e**

**4**

The finished bone inkwell in use with black walnut ink and a quill pen.

## 4: Make the top cork.

Repeat the process for making the base plate cork, except don't glue this one in or else your inkwell will become an airwell that you do not have access to. Also, don't saw the cork stick off flush with the bone this time. If you do that, you won't be able to grip your cork, and your inkwell will be stuck with a cork that can't be removed—another airwell situation. Instead, leave at least 1 inch (2.5 cm) or more of length on your cork stick so you can easily get a hold on it when opening your awesome new bone inkwell.

Inkwells and Paint Dishes

# 5 PAINTBRUSHES

The paintbrush seems to be the iconic symbol of the artist. Many people form an intimate bond to their paintbrushes, a relationship that deepens with time and experience. I still use and adore paintbrushes that I bought as an adolescent.

People have been using paintbrushes to apply pigments to surfaces since the Stone Age. These old-school brushes seem to have been quite disposable compared to the ones used by the masters of the Renaissance. This chapter explores a range of approaches of making artist paintbrushes, from rustic to more refined.

An assortment of rustic paintbrushes made from a variety of fibers

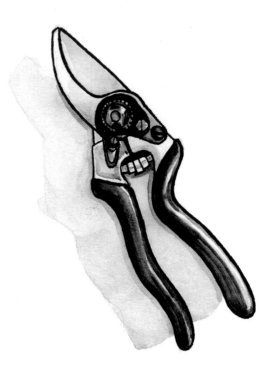

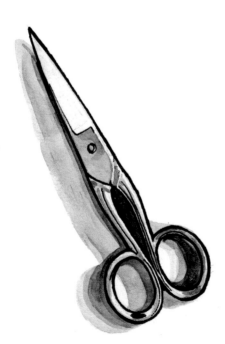

Tools used for making paintbrushes: knife, hand pruners, scissors, and tweezers

# Gnawed Twig Brushes

You can chew twigs from fibrous wood trees into paintbrushes of surprisingly high quality. I stumbled upon this discovery as a result of using twigs to clean my teeth on wilderness survival trips. In these situations, the twig begins as a tooth pick and by the time your teeth are clean it is a replica of a Stone Age tooth brush. It is a simple leap to apply this method of manufacture to make paintbrushes. In fact, it was not until after I had made this discovery that I learned of paintbrushes made in the early Stone Age in this manner.

You can chew twigs into surprisingly effective paintbrushes. Look for twigs of nonpoisonous, fibrous wood species. If it's hard to break a dry twig into two pieces, it may be a good candidate for a gnawed twig brush.

## MATERIALS

Twig from: beech (*Fagus* spp.), yellow birch (*Betula alleghaniensis*), sweet birch (*Betula lenta*), elm (*Ulmus* spp.), oak (*Quercus* spp.), or any other fibrous wood species

## TOOLS

Sharp knife
Hand pruners
Teeth or dentures
Utility knife or scissors

**1:** Find the right twig.

Look for a twig that is relatively straight and 8 to 10 inches (20.5 to 25.5 cm) long. It needs to be green, that is, a living twig from a live tree. If the twig is dead and dry, the wood fibers will be too brittle, and it will not make a good paintbrush. When you cut the twig from the main branch, be sure to make a clean cut from where the twig originates. This will help the tree heal faster and is a good way to say thanks for its gift.

**TIP:** When using a knife as a scraper, be careful to keep the sharp side of the blade oriented away from you.

**2:** Remove some bark.

Use your knife to make a shallow incision around the circumference of the twig, about 2 to 3 inches (5 to 7.5 cm) from the thicker end of the twig. Now use the back of the knife blade as a scraper to remove the bark down to the sapwood.

### 3: Shape the ends.

Whittle the thicker end of the twig into a symmetrical shape. This shape will designate the profile of the bristle bundle on your paintbrush.

### 4: Begin gnawing.

You may be surprised to learn that there is a fine art to gnawing. The basic principle that informs the technique is this: When wood fibers are dry, they are brittle and fragile; when they are wet, they are pliant and resilient. Therefore, it's important to begin by simply sucking on the debarked end of the twig. Hold it in your mouth as if it were an oversized toothpick. After a few minutes, begin to gently squeeze the end of the twig with your molars. Rotate the twig and squeeze it from this orientation. Continue this procedure slowly and patiently. If you bite down too hard at this stage, you will surely break the wood fibers and diminish the potential quality of your paintbrush.

### 5: Continue gnawing.

When the end of the twig begins to feel spongy or bouncy, you can begin to apply slightly more pressure as you squeeze with your teeth. Keep in mind that the wood fibers in the center of the twig are still dry and vulnerable. As you continue to gently gnaw, a hydraulic phenomenon occurs. Saliva is pushed into the wood when you bite down,

helping separate the fibers from one another. If you approach this process gradually, you will successfully separate the fibers into consistent-size fragments throughout the end of the twig.

### 6: Trim the bristles.

Once the center of the twig tip is rendered into fibers, you're finished gnawing. Now you may decide to trim the end of the brush to meet your fancy. Use a utility knife or scissors to make any modifications to the bristles, giving it a symmetrical profile.

### USE AND MAINTENANCE OF THE GNAWED TWIG PAINTBRUSH

Use this paintbrush as you would any other. You have permission to explore and push the boundaries of what is possible. When you're finished for the day, clean it as you would any other paintbrush, and then oil it to keep the bristles supple and resilient.

You can use any oil, although oils that don' go rancid are preferable. Consider walnut oil, tung oil, or mineral oil.

Paintbrushes made from plant fibers can be reverse-wrapped and tied together. The stiff fibers of agave and wild grape vine are shown here from top to bottom.

# Tied Bundle Brushes

Another simple way to make a paintbrush is to tie the bristle material directly to the handle. You can use this method to make paintbrushes from an endless selection of materials, but for this project you'll use pine needles. The pines trees (*Pinus* genus) are among the most widespread around the world, and therefore a pine species is most likely a member of your bioregion. The needles will be the bristles, and the straight twig will be the handle. These brushes won't withstand years of use, but they're easy and fun to make so don't worry about it.

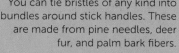
You can tie bristles of any kind into bundles around stick handles. These are made from pine needles, deer fur, and palm bark fibers.

## MATERIALS

Pine twig with needles
String
Waterproof glue (optional)

## TOOLS

Sharp knife
Hand pruners
Tweezers

**TIP:** For a super simple improvised paintbrush, clip a pine twig with a cluster of needles at its apex and use it as is. These have obvious limitations as well as less obvious rewards. Have fun with it!

## 1: Find a pine tree.

Find an actual pine tree with long needles, not a spruce, hemlock, or fir. These species are certainly worth experimenting with, but their needles are very short and not quite suitable for this project. Look for a stand of young trees that are growing in a dense cluster and select one that seems less healthy than its competing neighbors. Snip off a straight twig, usually the leading shoot at the top of the tree, or harvest the whole thing. Because of your interaction with the forest in this way, the other trees in the immediate area will have a better chance of reaching maturity.

**TIP:** Make charcoal sticks with the parts of the tree you don't use for the paintbrush.

## 2: Arrange a bundle of needles.

Remove the needles from the pine branch by pinching the base of each cluster. Next lay out 4 to 6 inches (10 to 15 cm) of string in a straight line on your work surface. Select your favorite needle clusters and place them perpendicular to, and on top of, the string, all facing the same direction. When you arrange enough needles for the size you want, tie them into a bundle with the string. Now you can shape the profile of the bundle by lining up the tapered ends flush, or to a pointed center.

tight and finish it with enough extra string to attach it to the handle with a secure knot. If the bundle can rotate around the handle, undo the lash and make it tighter this time, or use waterproof wood glue for a more secure bond.

## MAINTAINING TIED BUNDLE BRUSHES

As with all paintbrushes, let your tied bundle brushes dry out thoroughly before you store them. Moisture invites mold, and that can be a big problem for many reasons.

### 3: Prepare the handle.

Remove the bark from the last 2 to 3 inches (5 to 7.5 cm) of the larger end of the handle (from the base of the twig) and carve a short tapered point at its tip.

### 4: Join the bristles to the handle.

Insert the pointed end of the handle into the bristle bundle, keeping it centered within the needle mass. Now use the knotless lash (page 25) to connect the needle bristles to the twig handle. After the first few wraps, check the alignment of the tapered ends of the bristles to make sure there aren't any errant needles. Use tweezers to make adjustments at this point before they get squeezed too tight to do this easily. Make the lash nice and

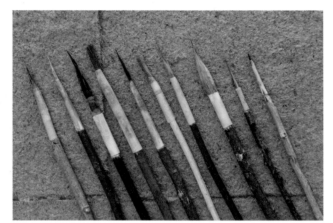

An assortment of animal fiber paintbrushes. These paintbrushes use a ferrule made from a turkey feather to attach the bristles to the handle. The finest paintbrushes are made from hair, fur, and whiskers.

*Looking Out.* Nick Neddo 2014, pine soot ink wash with handmade paintbrushes made with found cat whiskers

**TIP:** Oil the needle bristles to keep them supple and prolong their longevity. Walnut oil, tung oil, and mineral oil all work well for this.

# Brushes with Ferrules

These paintbrushes fit a spectrum of slightly more advanced to much more advanced, depending on the materials used and the level of persnickety-ness and patience of the brush maker. Fine paintbrush making is an art unto itself, full of secret techniques and subtle nuances and tricks. To achieve the highest levels of mastery in this art requires lifetimes of skill and experience, usually passed down from master to apprentice. However, just as with any skill, with each paintbrush you make and then use, you deepen your understanding and relationship with the individual tool as well as the overall art. You'll find that even your first few paintbrushes will have their place in your work; their creative potential and technical versatility are directly related to your imagination and sense of exploration.

You can obtain fur from fuzzy creatures without causing them harm. Firefly, a Nigerian dwarf goat, helped make a paintbrush or two.

## MATERIALS

Turkey feather
Thin, straight, strong stick
Fur, hair, or whiskers
Waterproof wood glue
Liquid cornstarch
    (optional)
Strong thread or small
    rubber bands

## TOOLS

Sharp knife
Paper clip or tooth pick
    (optional)
Sandpaper (optional)
Clean, white sheet of paper
Tweezers
Magnifying lens
Ruler
Sharp scissors or a utility
    knife

## 1: Get some fur, hair, or whiskers.

Before I go any further, you must know that it's not necessary or condoned to harm or kill any animals for this project!

Several times each year, animals shed their seasonal fur, leaving it everywhere they hang out. If you have a pet with fur, it's easy enough to start collecting it; otherwise, you can look to the pets of friends. You can buy old fur garments in secondhand stores or at estate sales for cheap. Find a garment that's in a state of disrepair. These are great candidates for repurposing into a new entity with a new life. Often I find tufts of fur at kill sites when

I am tracking wildlife on the landscape. Consider this a lucky offering from the sacred dance of predator and prey.

If you're a hunter, of course, you might have access to fur in the event that you bring fresh meat home to your family.

If you want to use hair for the bristles of your paint bush, consider a lock of your own or one from a friend. Horse hair is commonly available from horse farmers. The same is true for pigs, which provide the bristles for many fine commercial paintbrushes.

Whiskers are highly prized bristle materials. One reason for this is due to the structure of the whisker itself. They are stiff yet flexible, and like fur they gradually taper to fine points at their tips. Another reason whiskers are desired by the brush maker is because of their lack of accessibility. Please, don't cut whiskers off from your pets! It is cruel and unnecessary because they rely on their whiskers for proprioception and sensing subtle nuances in their environments. Instead, consider the whisker to be a found treasure. Aside from dead animals, most of the whiskers that I have found were gifts from my cats. I found them on the floor when I was sweeping. I admit that when I snuggle with my cats I occasionally check to see if there are any loose whiskers by giving them an affectionate tug. I have yet to procure one in this way.

Nevertheless, it's best to save the whiskers until you have more experience making finer paintbrushes. This is convenient because it takes a while to scavenge enough for a small brush anyway.

## 2: Prepare the ferrule.

The ferrule of the paintbrush is the tapered tube that holds and connects the bristles to the handle. The base ends of turkey feather quills have been used to make ferrules for centuries, and this is what you'll use for this project. Choose a quill based on its size relative to the size of paintbrush you want to make: larger quills for larger brushes and so on. Remove the barbs of the feather as you would if you were making a quill pen. (See "Quill Pens" on page 36.) Use your thumbnail or the back of a knife blade to scrape off any membranous tissue from the end of the quill. Next, use a utility knife with a sharp blade to score the quill about 1½ inches (4 cm) from the end—a bit shorter for a smaller paintbrush; longer for a larger one. Continue scoring all the way around the quill until the ferrule separates from the rest of the quill shaft, being careful not to crush or flatten it as you do so.

Use a paper clip, toothpick, or the top end of the quill to clean out the inside of the ferrule. A thin layer of cellular tissue will come out with a little effort. Make sure the hole at the tip of the ferrule is open. Push the paper clip or toothpick through and remove any remnant membrane while you are at it. If the tool you use to ream through is too big it will split the end of the ferrule. Proceed with caution.

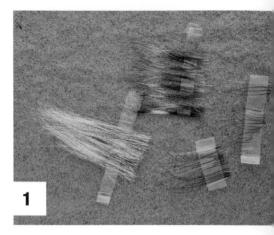

Hold the precious fur and whiskers that you collect over the years in a piece of tape for easy location and storage. Learn by making brushes with fur and hair first to build your skill and craftsmanship before considering using whiskers.

Score and cut the base of the feather 1 to 1½ inches (2.5 to 4 cm) long. Cut the tip off, leaving a diameter half that of the other end. Use the tip of the feather remove the membrane inside the ferrule.

## 3: Make the handle.

Use a strong, straight, slender stick, about 8 to 12 inches (20.5 to 30.5 cm) long, depending on your preference. The base end, which is larger in diameter, will be the brush end. This end needs to be whittled to be small enough in diameter for it to fit at least ¼ inch (6 mm) inside the ferrule. The taper at which you carve this end should also match that of the inside dimensions of the ferrule, making a custom match. Carve the other end of the handle to a clean edge. Sand it smooth if you wish.

## 4: Arrange the bristles into a bundle.

If you intend to use the natural taper of each individual fur and/or whisker to create the overall shape of your paintbrush, this part is difficult. However, if you're using hair (which has no taper) or you intend to trim the bristles to shape later (disregarding the natural taper of fur), this step is fairly simple. This is one of the biggest distinctions between a nice brush and a very fine brush. The following instructions proceed with a technique for using the natural taper of the fur.

First, lay out about 8 to 10 inches (20.5 to 25.5 cm) of thread onto a clean, white sheet of paper. This will be used to tie the bristles into a bundle later. Next, select individual strands that match in length, taper, and bend. Place these onto the thread with the tapered ends lined up. When you have as much as you need for

the size brush you want to make, tie them together into a bundle with the thread. Use tweezers and maybe even a magnifying lens to pull individual strands into position. They should either all be side-by-side or shortest on the outside diameter of the bundle and longest in the center. This takes diligence and patience, as well as a steady hand. This is also the reason why there are not a lot of master paintbrush makers out there.

**4a**

Use a white piece of paper and a pair of tweezers to arrange individual fur pieces over a taught length of thread. This is tedious work that requires a steady hand and a lot of patience.

**4b**

Arrange the bristles into bundles and tie them together with thread. The tapered ends of each strand of fur should line up according to the contour desired for the paintbrush.

**TIP:** You can use small rubber bands to hold the bundle together in place of thread.

## 5: Insert the bristle bundle into the quill ferrule.

Measure the overall length of the bundle to leave ¼ to ½ inch (6 mm to 1.3 cm) of space in the base of the ferrule when it's inserted. Make sure your knot is tight, and then use sharp scissors or a utility knife to trim the base ends of the bristles with a straight, clean cut. Make adjustments to individual strands as needed and then insert the tip of the bundle into the ferrule's larger opening. Use a small twig or toothpick to push the base end of the bristle bundle through the ferrule until the bristles protrude from the smaller opening at the length you desire for your brush.

When the bristles are successfully fitted into the ferrule, it's time to glue them together at the base. Put a small dab of waterproof glue into the larger opening of the ferrule where the base ends of the bristles are exposed. Let it sit overnight in a safe place, away from pets and children.

## 6: Attach the handle.

Without being too forceful, test the fit of the handle into the ferrule. If it's too tight, whittle it down to make a precise, snug fit. If it's too loose, trim it and prepare another section of the stick to fit. Put a dab of waterproof wood glue into the ferrule and join the two pieces and wipe off any excess as it overflows the seam.

Use tweezers to pluck out any strands that may be causing trouble.

**TIP:** Consider using a dab of glue or liquid cornstarch on the base of the bundle to facilitate step 5.

Trim the base ends of the bristle bundle even with one another before inserting it into the quill ferrule. Leave ¼ to ½ inch (6 mm to 1.3 cm) of space at the wide end of the ferrule.

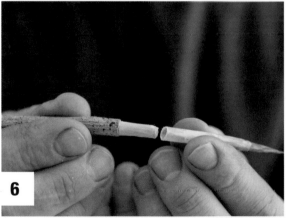

Put a few drops of waterproof wood glue into the ferrule's opening and insert the whittled end of the handle where you left space when positioning the bristles. Do not force the fit or the ferrule may split.

The finished paintbrush in use.

*Misty Morning.* Nick Neddo 2013, goat hair paintbrush and pine soot ink wash

# 6 | PIGMENTS AND PAINTS

A collection of stones and the natural earth pigments and paints made from them

Pigments are the elemental components of color. They're the source of vibrancy to every color we see, thanks to light and the cones and rods in our eyes. There are several sources of pigment that can be broken down into four categories: plant-derived, mineral-derived, fossil fuel–derived, and synthetic. Let's take a closer look at each of these.

# PLANT-DERIVED PIGMENTS

The plant world is full of color. Every flower, leaf, and root is bursting with pigment. The only catch is that these pigments are not stable by themselves. Over time, they decompose and change as they age, just as a brilliant red autumn leaf turns a tannish brown on its way to becoming soil again. The abundance of green on the landscape comes from chlorophyll, which fades within hours from being extracted by the plants. Therefore, plant pigments require some additional chemistry to make them lightfast. A mordant such as alum is needed, and this gets into the whole world of natural dying, which is unfortunately beyond the scope of this book.

# MINERAL-DERIVED PIGMENTS

Likely most of the colors used in ancient Paleolithic artwork found on the walls of caves, body paint, tattoos, and on the remains of the deceased, were made from rocks. These pigments are known as natural earths, and they're made by crushing and grinding stones into a fine powder. The earliest known use of these pigments dates back to the Paleolithic period (350,000 BCE), and they've been in constant use in every civilization since.

A considerable amount of effort is required to process natural earths, depending on the hardness of the parent rock being used. This offers a bit of an insight into the world of our Stone Age ancestors. Art must have been important to them, otherwise the time and energy spent on grinding rocks into pigment wouldn't have been tolerated or sustainable.

These pigments are lightfast. Each bioregion on Earth offers its residents its own unique color palette, based on the local geology. Essentially, the amount of colors available in an area match the variety of colors reflected in that area's stones. It's fun to imagine a world where our homes blend into the landscape, painted with the colors from the ground where they stand.

# SYNTHETIC PIGMENTS

The first synthetic pigments were created by the ancient Egyptians, sometime around the third millennium BCE. This was a blue pigment (a double silicate of copper and calcium) born of fire. The manufacturing of synthetic pigment was an unrivaled innovation in the ancient world, and it was promptly exported to the Romans who named it "Alexandrian blue." Blue earth pigments are rare and expensive, which explains why they are scarce in the archeological record, and why synthetic blues were adopted with great enthusiasm.

The endless variety of colors available these days is easy to take for granted. It wasn't until the eighteenth century when chemists started making breakthroughs in synthetic pigment production. Today the vast majority of pigments are made from chemicals derived from fossil fuels. A quick walk through a paint store reveals that industrial chemistry has made the world pigment rich in a way that our Stone Age ancestors would likely have found difficult to grasp.

# Paints and Binders

Paint is not just one thing. What all paints of a specific color have in common is the pigment that gives it that certain color. It's the added binder that determines what kind of paint you have. Let's take a closer look at binders, shall we?

Binders, or vehicles, are the defining ingredient that distinguishes one type of paint from another. The binder keeps the pigment particles in suspension within the medium. Left to right: honey, eggs, oil, hide glue, gum arabic, and tallow (rendered fat)

## GUM ARABIC

This substance is derived from the sap (or pitch) of a couple members of the acacia trees, including *Senegalia senegal*. This tree resin has been used for centuries as a binder in watercolor paints and inks, and it's prized for its ability to keep pigments in suspension within a liquid medium. I have not included gum arabic as a binder in the projects within this book because I have not yet been to the Sahel region of Africa where these acacia trees grow. However, it's widely available on the worldwide market, and it has been for many centuries. Lucky for me, plenty of other substances are also ideally suited as binders.

## HIDE GLUE FOR WATERCOLORS

Hide glue is the binder of choice for me because it's widely available on the landscape (provided you can get your hands on a hide), and like gum arabic, it's water soluble but hard and stable when it's dry. These properties make for an ideal binder in watercolor paints. (See "Make Hide Glue" on page 65.)

Watercolors are amazingly potent paints that wait dormant until someone (you) wakes them up by splashing them with water. Once awakened (some would say rehydrated) the pigments can be gathered up by the bristles of a paintbrush and then dispersed in all manner of creative fashions.

## SALIVA FOR GENERAL PAINTS

Ready? Put some of your powdered pigment into a small container such as a mussel shell or small jar. Now add saliva to it and mix it to the consistency that you want to work with. That's right, spit and stir.

This might sound distasteful to some and repulsive to others, but you should know that saliva is quite well suited to be a binder. I suspect that saliva is the original binder (aside from water), and there's some archaeological evidence that supports my theory. It's believed that much of the ancient paint found on the walls of caves throughout the world was made with human saliva as the binder. Saliva is thicker than water, and it has a complex cocktail of enzymes that happen to be perfect for mixing with powdered stones to make paint. This is one of my favorite binders for making simple paint in the field, and for use on stone surfaces as the support.

## OIL AND TALLOW FOR OIL PAINT

Use tallow (rendered animal fats) or any oil that is not prone to rancidity as your binder, adding more for thinner paint and less for thicker paint. Experiment with different oils and quantities to get the consistency you're after.

## EGG WHITES FOR TEMPERA PAINT

The whites of eggs are the binder for tempera paint. Add powdered pigment to the egg whites in quantities that give you the color density you like.

## HONEY AND PRINTMAKING PAINT

At this point in my level of experience, honey seems to be the best locally available binder to add to pigments for creating a thick, tacky ink suitable for printmaking. The trick is to keep the ink thick enough so it doesn't mold on the paper yet thin enough so it can dry and not be sticky.

# Mineral Pigments and Paints

These are the colors available from the rocks you find on the ground. They all seem to go well together and offer unlimited creative possibilities. The following project is a prerequisite for making crayons. Once you have processed a pigment, you can use it to make a bunch of different art mediums, including oil paint, tempera paint, watercolor paint, crayons, even ink. As with any handmade process, this will change the way you look at rocks. For some people, a rock is just a rock. For others, it's a decoration or a tool. After you make this project, a rock will be paint as well.

You can crush or grind rocks of almost any variety down into pigment. Sedimentary rocks are by far the most willing, although fine colors come from harder igneous and metamorphic stones as well.

## MATERIALS

Rocks
Water
Binder of your choice (See "Paints and Binders" on page 106.)

## TOOLS

Stone mortar and pestle
Rock hammer
Safety glasses
Clear glass jars with tightly fitting lids
Canning funnel
Fine mesh filter (Tea strainers work well.)

Tools needed for making pigments and paints: stone mortar and pestle, canning funnel, fine mesh tea strainers, and jars

## 1: Find some stones.

Go to a place where you can source a wide variety of stones. River and stream banks and gravel bars are excellent places to find rocks of many types and from a variety of origins. Sedimentary rocks are by far the easiest to grind into powder, and they're the source of most of my mineral pigments.

## 2: Test some rocks.

All stones can provide you with pigments. The question is how hard you're willing to work for them. Some stones seem to give them to you, while others make you earn them.

When you find a stone that interests you, try drawing with it on a larger rock, as if it were chalk. If the rock that you are drawing on is harder than the rock you are testing, the residue left behind is pigment from the test rock. Some rocks will draw easily leaving bold lines, while others leave more subtle traces of color behind. Some rocks, such as the metamorphic family, are so reluctant to share their pigments that scarcely a trace of color is left, with only a scratch-like engraving remaining. Try making lines on several different rocks to get a better sense of the stone you are testing. Gather up what you want to experiment with and bring them back to your studio space.

## 3: Break up rocks.

If your rocks are larger than a table-tennis ball, you will need to break them up. Break large rocks into smaller pieces with a rock hammer rather than trying to do it in the mortar and pestle. I have broken high quality granite mortars by trying to bust up large stones in them. Wear safety glasses or other eye protection when you are crushing rocks. It's a bummer to get rock fragments in your eyes!

Strike the stone from above with light force at first. If the rock does not break apart after several blows with light force, proceed by increasing the power of your impact as you continue to strike the stone.

## 4: Grind.

Get out the stone mortar and pestle and place a stone that's smaller than a table-tennis ball in it. Use an alternating grinding and crushing approach. Grind by applying moderate downward pressure as you circle the bowl of the mortar from inside center to outside, and back again. Now focus on the largest chunks in the mortar and crush them with the same striking action that you used to make the initial fracture. Continue to crush and grind until the rock powder appears to be consistent in texture.

Use a stone mortar and pestle to crush and grind rocks into pigment. The finer the particulates, the better quality your pigments will be.

**TIP:** Work with small batches of stone at a time when you grind pigments. This will help you be more thorough and avoid accidental waste.

## 5: Sift.

Set up a glass jar with a fine mesh filter fitted into the mouth. Place a canning funnel inside the fine mesh filter and carefully pour the rock dust into it, a little at a time. Agitate the funnel while keeping it above the filter so you don't lose any precious pigment. The fine particles will filter through, leaving the larger bits behind. Repeat this with all of the rock dust you made and return the larger particles back to the mortar and pestle to finish grinding them. Repeat this process until all of the particles make it through the filter. At this point, you have a rustic pigment that can be used for all sorts of things. Or you can take this a couple steps further, improving the final product significantly.

## 6: Grade.

The smaller the rock particles are, the finer the quality of the pigment will be—and thus everything you make with it. When you're satisfied with the level of fineness of your pigment, you can begin the water grading process. This is a simple, yet ingenious method for purifying pigments that people have been using since the Stone Age. First, add a little water to the jar that you just filtered the rock particles into. Go for about twice as much water as pigment. Put a water-tight lid on and shake the mixture vigorously for about a minute. Seconds after you stop agitating the pigment slurry, you'll see the larger, heavier particles settle to the bottom while the smaller, more buoyant particles remain suspended in the water. Pour this fine pigmented concoction into a small jar (like a mini jelly or jam jar) slowly, being careful not to let any of the sludge at the bottom in. The pigment sludge that didn't make the cut can be used as a more rustic paint, or it can be dried and reprocessed in the mortar and pestle later. The new jar of fine pigment is the super nice stuff and should be set aside somewhere safe to evaporate into a more concentrated paint base or back into a powder.

## 7: Make Paint!

Turn your pigments into paint by adding a binder. A good place to start is with one part binder to one part pigment. Mix them together until the paint has an even texture and consistency, without any lumps or dry pigment. If your binder is hide glue, prepare it as described in the inks chapter (page 49). For powdered gum arabic, mix one part powder into two parts warm water and stir until fully dissolved before adding to your pigments. Whether you use oil, tallow, lard, egg whites, hide glue, gum arabic or saliva, experiment with the binder/pigment ratio to get the consistency that you like.

**5**

Pour the rock powder into a jar through a fine mesh filter, such as a metal tea strainer. Agitate until all that remains is the larger particles. Put these back in the mortar and grind them finer with the pestle.

**6a**

On the left is a pile of coarse, rustic pigment, and on the right is a pile of highly refined, purified pigment made from the same stone. The former is well suited for working in the field; the latter of a higher quality, for the studio.

## 8: Store.

When the pigment is dry, you can put it away for future use. Put a small piece of the parent rock into the jar with the pigment to keep in touch with where it came from. This will help you stay organized if you want to replenish specific colors or when you have too many pigment jars to keep track of by memory.

**6b**

Put the ground rock powder into a jar and add a small amount of water. Put the lid on and shake vigorously for a few seconds. The finer particulates will stay in suspension; the larger ones settle to the bottom.

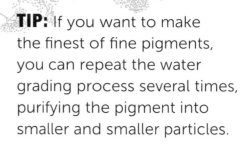

**TIP:** If you want to make the finest of fine pigments, you can repeat the water grading process several times, purifying the pigment into smaller and smaller particles.

**6c**

**7**

Prepare and mix in the binder of your choice, depending on the type of paint you wish to make.

**8**

Pigments made from stones can be organized and stored in small jars. Consider leaving a small portion of parent rock in the jar with the pigment for identification purposes.

The author uses his handmade paints on a rawhide paint box.

# 7 CRAYONS

Crayons were the first artistic medium that I got my hands on early in life. Later, as I began to explore and prefer pens, pencils, and markers as drawing tools, the crayon box began to collect dust. Over time, my relationship with crayons grew stagnant and fell into the realm of something that only children use.

Don't let the crayon's association with children's art allow you to overlook the creative potentials of this unique drawing tool! Making my own crayons has launched me into a new phase of creative exploration. It's almost as if I am meeting an old friend or family member again after years of separation. There's something about having a crayon in my hand that connects me back to the playfulness of childhood. It seems that simply holding a crayon with the intent to make a drawing gives me permission to be playful, to relax, and not take myself too seriously. This is noteworthy because when we lose this playful approach to making art, we tend to stop making it.

An arrangement of crayons handmade from beeswax, stones, charcoal, and soot

# CRAYON MOLDS

The mold is as important as any other ingredient of the crayon. You need a mold to give your crayons form, and what you use will be recorded in the final form and shape of your crayon.

A simple, yet decidedly not available from the living landscape solution, is the big gulp variety of plastic straws. These can often be "foraged" from your local convenience store or fast-food restaurant. Just tape up one end of the straw to make a seal, and you have a crayon mold. When you're ready to pour crayons, set up your straw molds into your mold holder. After the crayons have cooled and hardened, carefully make an incision along the length of the straw to get your crayon out. Convenient, yes; inspiring, perhaps not.

The most exciting and field-expedient crayon molds that I've used are made from the stalks of Japanese knotweed, which I discuss in more detail a bit later.

Be creative and explore your bioregion to find other crayon mold options.

## WAX

Wax is the binder used for making crayons. Of the wax options to choose from, beeswax is the most interesting because it's manufactured by bees using flowers as their raw materials. Paraffin wax is, in some ways, more ideally suited for making crayons. It's less tacky to the touch than beeswax and, more importantly, it's white, which doesn't interfere with the color of your pigment. On the other hand, paraffin has a major downside, which is that it is made from fossil fuel oil, which makes it less available to those of us who want to make our tools by hand. The following recipe will work well with either beeswax or paraffin wax. You can decide what kind of crayons you want to work with.

Beeswax in its many forms, from left to right: raw honeycomb beeswax, rendered beeswax, and chopped beeswax

Tools used in this chapter: small pot, funnel, knife, teaspoon, campfire, saw, hot plate, and scale for the adventurous

# Waxy Drawing Sticks

The colored crayons you make will come from stones (mineral pigments, also called natural earth pigments), whereas the black crayons will use pigments from charcoal or soot.

## MINERAL PIGMENT CRAYONS

The color for these crayons is from rocks. The name of the game is to smash, crush, pulverize, and powder a stone until it is as fine as dust. This is called mineral-derived pigment. (See "Mineral Pigments and Paints" on page 108.)

## CHARCOAL AND SOOT PIGMENT CRAYONS

You can make black crayons by using powdered charcoal or soot (referred to as lampblack) as the pigment (as in the photo sequence). You can collect charcoal from a campfire or your charcoal kiln. (See "Charcoal" on page 15.) Grind these pieces of charcoal into a fine powder with a ceramic mortar and pestle to make pigment. The blackest crayons come from soot pigment. Make your own (see "Inks" on page 49) or get some lampblack from the wide world of commerce. Take extra effort to mix the soot in thoroughly with the wax when you add it in.

*Bee People.* Nick Neddo 2013, beeswax crayons pigmented with stones and charcoal

## MATERIALS

Powdered pigments from stones, charcoal, or soot
Japanese knotweed stalks
Beeswax

## TOOLS

Folding knife or fine-toothed saw
Cardboard box
Small pot
Teaspoon
Stirring spoon
Hotplate or kitchen stove
Small funnel
Paper napkins or paper towels
Digital scale (optional)

## 1: Process the pigments.

The color of your crayons will come from the pigments you make from stones, charcoal, or soot. Refer to the "Pigments and Paints" chapter (page 103) for instructions on how to process pigments from rocks. To make soot for a deep black pigment, see the "Inks" chapter (page 49).

1

Use stone pigments or grind charcoal in a mortar and pestle to produce a black pigment. When the pigment is finely powdered, it's ready for storage or to be mixed in with the beeswax.

## 2: Harvest the molds.

You can make crayon molds from a variety of hollow things that match the diameter of the crayons that you want to make. For this project, I would like to introduce Japanese knotweed, whose stalks work perfectly. They are much like bamboo in the way that they grow a stalk with hollow chambers divided by nodes.

A notable difference between Japanese knotweed and bamboo is that knotweed has much less structural integrity. This weakness is actually a strength for the crayon maker because it allows you to easily peel away the woody walls of the stock when revealing the crayon from the mold.

The epitome of invasive species, Japanese knotweed is increasingly found along waterways and edges throughout the world as it extends it range into new territories. Because it's here to stay, we might as well learn to incorporate it into our material culture.

Summer growth works well, but I prefer to find knotweed in the fall or early winter (before it breaks down) and harvest the dead plant stalks from the most recent growing season. Reject any stalks that have cracks in them. These are too old and often have a more faded, sun-bleached color than the burgundy-color of the most recent, dry stalks. Use a folding knife or fine-toothed saw and harvest these dead stalks toward the base.

## Important!

Be sure to leave any seeds or seed cluster residue there in the knotweed patch so you don't inadvertently spread this invasive species faster than its managing to do itself.

## 3: Make a mold holder.

A cardboard box works well as a simple mold holder to support the molds while you're pouring hot pigmented wax into them. Get ahold of a small cardboard box and flip it upside down so the bottom is facing up. Using a knife, make rows of x-shaped slits 1 or 2 inches (2.5 to 5 cm) apart from one another. With the bottom of the box full of these perforations, you're ready to move forward. You can also make a sturdier mold holder out of wood, with varying-sized holes drilled into the top to fit an assortment of molds.

## 4: Prepare the molds.

Cut the knotweed stalks into small segments of at least ½ to 1 inch (1.3 to 2.5 cm) in diameter, with a node at the base of each piece. The node is going to be the bottom (or the floor) of the crayon mold and needs to be water tight. If you see damage to the node, discard that segment and move on to another one. With the molds sorted and processed, it's time to insert them into the mold holder. Simply push each mold into an x-hole so that it is held into an upright position. Fill the mold holder with as many molds as you plan to use for the crayon batch you have in mind.

**2**

The author gathering Japanese knotweed (*Polygonum cuspidatum*) stalks. Although considered an aggressively invasive species, their brittle, woody stalks are ideal for use as crayon molds. Process on site: Remove branches and leave them there, taking only the stalk's main trunk.

**4**

Knotweed cut into crayon mold lengths

## 5: Heat the beeswax.

Use a knife to cut a lump of beeswax into small pieces to help it liquefy faster in the pot. Measure 3 or 4 tablespoons (45 to 60 g) of the chopped wax and put it into a small pot on low heat. It's best to bring the beeswax to a liquid state slowly to keep it from burning, which makes a lot of smoke and darkens the wax. Stir the wax regularly as it heats up and liquefies.

## 6: Add the pigment.

There's a lot of variations within the recipe, depending on how dense you prefer the pigment to be in your crayons. The more pigment you add to the wax, the more intense and saturated the color of the crayon will be. There's a point, though, where too much pigment in the crayon will make it weak and likely to break while drawing. Keep notes on your pigment/beeswax ratio when you are experimenting.

To begin with, measure 1 to 3 teaspoons (4 to 12 g) of pigment for each medium-size crayon. Add it to the liquid wax, stirring constantly and thoroughly. Continue mixing the pigment into the wax until there are no clumps and the whole concoction becomes consistent in texture.

**TIP:** Hot wax is dangerous, and it can cause serious burns if it comes in contact with skin. Be cautious and deliberate when handling hot wax!

## 7: Fill the molds.

Fit a small funnel into the opening of the crayon mold that you want to pour into first. Give the pigmented wax another good stir just before you start to pour it into the mold. This helps to distribute the pigment more evenly among the wax solution. Slowly and deliberately pour the colored wax through the funnel and into the mold. Pour carefully so you don't overfill the mold. With experience, you'll be able to fill each mold to the brim without spilling a single drop of the precious crayon wax. As the hot wax fills the space of the knotweed shaft, the wax level will go down slightly. If you wish to fill it up to the top, be sure to do it

before the wax from the first pour begins to harden. Next, allow your crayon to harden in its mold for at least two hours before attempting to remove it.

The best time to clean out the crayon pot is before it cools down after pouring the crayons. Use a paper towel or napkin to absorb the pigmented waxy residue that you couldn't salvage. This only works if the pot is hot enough for the wax to remain in a liquid state. Return the pot to the hot plate as needed and be careful not to burn yourself on the hot pot. Use a stick to mop the paper towel around the inside of the pot if you need to.

**TIP:** If you want to get more consistent results with your crayon recipes, use a digital scale to measure the wax and pigment by the gram. Find the ratio that works for what you are looking for.

**5**

Use a small pot on a hot plate to melt the beeswax. Be sure to work in a well-ventilated area.

Add the pigment in with the liquid wax gradually while stirring it in.

**6a**

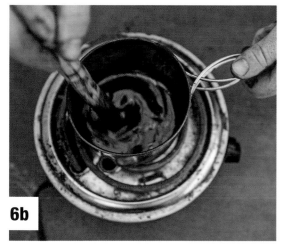

**6b**

Stir the pigment in thoroughly to avoid clumps of unmixed powder in the wax.

**7**

Use a small funnel to pour the pigmented wax into the molds. Fill them up to the top and then lightly tap them to help the air bubbles come to the surface before topping them off. Don't burn yourself!

## TIPS:

When pouring crayons, give yourself plenty of space to handle this potentially dangerous liquid. Clear away any unnecessary objects that might interfere with this step to help ensure your safety.

Agitate the molds lightly by tapping on them immediately after pouring the pigmented wax. This will help the air bubbles come to the surface before they get stuck in your crayon as a hollow spot.

Collect the stray bits of wax that spill off when you are filling the molds. You can melt down these loose bits to make new crayons, often with the result of a new brown when all mixed together.

## 8: Remove the crayon from the mold.

When your crayon has cooled and hardened, it's safe to peel away the Japanese knotweed mold that contains it. If you have any doubts regarding whether or not it's ready, wait another hour. You will know for sure when there is no detectable heat coming from them when you hold them in your hand. Knotweed usually comes off in quarter panels as it's peeled away from the crayon. Gently squeeze the mouth of the mold until it begins to crack. Rotate the mold a quarter turn and squeeze the mouth again from this orientation. Once the knotweed has cracked a bit, peel it away from the crayon inside, one panel at a time. Be patient and take extra care not to break the crayon as you peel the mold. This can be hard to do because opening a crayon is much like opening a present.

When the crayons have cooled and hardened in their molds, it's time to unwrap them. Japanese knotweed is fairly brittle, and it can be cracked and peeled away to reveal the crayon inside.

**TIP:** If you prefer, you can leave your crayon inside its Japanese knotweed shell and use it as a protective case. Just carve and peel away the woody shell when you need to sharpen it.

## Using Beeswax Crayons

Mineral pigment crayons are fun to work with because each color batch seems to have its own feel and behavior. Some crayons have a smooth, almost greasy action, while others are more gritty and abrasive. Some crayons leave subtle traces of color behind on the surface, and others leave brilliant and opaque marks.

Several factors influence these characteristics in the performance of a crayon. A few things to consider while experimenting are the inherent properties of the source rock, how coarse or fine the pigment was ground, the pigment/wax ratio, and how much pressure is used to make lines on a surface.

Crayons allow you their own unique creative opportunities and techniques for creating interesting images. Explore the medium by layering, scraping, scratching, rubbing, and using them in combination with other mediums.

Earth pigment crayons in use by the author

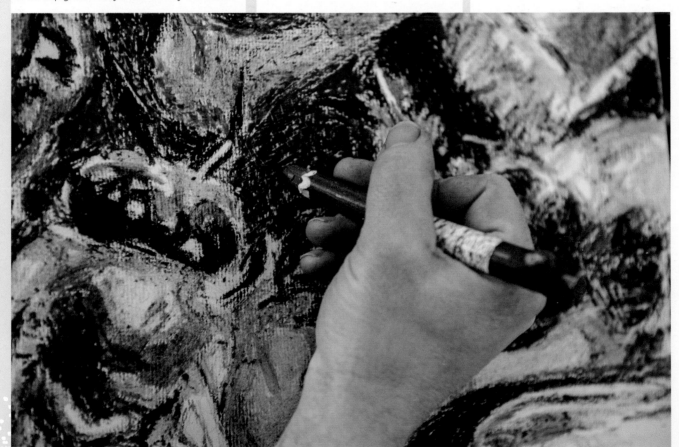

# 8

Handmade paper cascades out from a cattail basket, all made by the author, from right: wasp nest, sphagnum moss, cedar, wild grape, cattail, lawn clippings, milkweed, straw, cotton rags, willow, and eastern hemlock

# PAPER

Paper making is relatively simple in theory. The quality of paper you make can range from rustic to highly refined. The experience and skill of the papermaker are factors that play into the craftsmanship of the finished product, along with the tools and techniques used to make it. The goal of this chapter is to get you started making paper and building your paper-making studio. If you get serious about paper making, you may want to upgrade your tools as you see fit. I will get you started with simple tools and give you some options for upgrading if you want to take it to the next level.

## MOLDS

The mold is the tool that draws the paper from the vat of pulp. It's a rectangular frame, usually made of straight-grained hardwood to withstand the demands of the paper-making process. The molds are covered with a tightly stretched sheet of porous fabric, such as burlap, silk, linen, gauze, or window screen. This sheet of material holds the pulp onto the mold while allowing water to drain through it.

## DECKLES

The deckle fits tightly over the mold and prevents pulp from building up on its outer edges when forming a sheet of paper. Using a deckle is not entirely necessary, although it makes it easier to transfer (couch) the paper onto the felts later on. This the tool gives handmade paper its characteristic "deckled" edge.

## PAPER PRESS

The same press you might use in the printmaking and book-binding process can be used for making paper as well. The press removes excess moisture from the paper and compresses it flat.

## FELTS

When pressing paper, sheets of felt are stacked between each waterleaf (a newly formed piece of paper) and the surfaces of the press. These absorb water and

A wooden picture frame and some burlap can make a simple, yet effective mold.

also cushion the fresh sheets of paper during pressing. The felts need to be a couple of inches (5 cm) larger on all sides than the sheets of paper you make. You can use old wool blankets, commercial felt, or even cotton towels for this.

## BLENDER

After the pulp material is boiled, it gets blended into a pulp. In the old days, people did this by hand with wooden tools, and it took hours. With a blender, you can make pulp in minutes. As far as blenders go, you don't need anything fancy. The most reliable blenders seem to be the old ones that are made of metal and glass, rather than plastic. Keep your eyes open at thrift stores and flea markets.

## VATS

A waterproof vat or tub of some sort is essential for making paper. The vat is where the pulp is reconstituted and made ready for drawing paper with the mold. It needs to be larger than your mold by at least a few inches (7.5 cm) on all sides so both the mold and your hands can fit into the vat together.

# Make a Simple Mold

You can make a perfectly functional mold from a wooden picture frame and a small sheet of burlap. This mold can be used without a deckle.

Use a heavy duty staple gun to stretch the screen onto the frame, making it as taut as possible. Tap protruding staples in with a hammer and trim off the excess burlap with scissors.

## MATERIALS

Wood picture frame
Burlap

## TOOLS

Scissors
Heavy duty staple gun
Staples
Hammer

## 1: Collect the materials.

Go to a secondhand store or flea market to peruse the picture frame selection. Look for one that's within the size of your vat and that will make paper of the size that you want. For your first mold, I recommend keeping it on the smaller side. Find a picture frame that's in relatively good shape.

You can often find burlap at cafés for free or cheap. Coffee is shipped in large burlap sacks of varying mesh coarseness. Look for the finer mesh burlap if it's available. If not, the coarse burlap will do. Fabric stores sell burlap by the yard and may be your best option. Be sure to cut the burlap screen to be larger than your frame by at least 4 inches (10 cm)

on each side. You can trim off any excess burlap later.

## 2: Stretch the burlap over the frame.

Lay out a sheet of burlap on a flat surface. Now place the picture frame on top of the burlap and cut out a section that is larger than the frame by 4 inches (10 cm) on each side.

The screen will be stapled to the outside edges of the picture frame, keeping it pulled as taut as possible without damaging it. First, lay the burlap over the front face of the picture frame and put a staple in the center span of one

of the edges, pinning the screen down to the outside edge of the frame. Use a hammer to tap the staples in all the way if necessary. Pull light tension toward the opposite edge of the frame and put another staple in, directly across from the first. At this point the burlap screen is attached to the frame from two opposing points. Now staple one of the other two edges, again in the center of the span. Pull tension toward the opposing edge and pin it down. Staple one of the corners down, pull tension, and then the opposite corner. Continue attaching the screen to the frame by stapling in pairs of opposing tension until there are staples every ½ inch (1.3 cm) apart. The screen should be quite taught when you're finished. If it sags or has a lot of play, you should remove the staples and try again with more tension. When you're done, trim off the extra burlap.

A mold like this can make paper without the use of a deckle. The paper is formed on the mold and allowed to dry there, instead of couching it onto a felt blanket. When dry, the paper is carefully removed from the mold screen.

**TIP:** You can use fiberglass window screening in place of burlap for the mold screen.

## MOLD UPGRADE

You can make a more professional mold by using higher quality materials, adding handles, eliminating metal hardware that creates rust, adding a ribbed support structure, and replacing the burlap screen with a 30 mesh (600 micron) brass screen (30 strands of wire per inch [2.5 cm]).

Use a species of wood that can better tolerate the demands of being submerged in water, such as oak. Cut strips of ¾ inch (2 cm)-thick board to be 1½ to 2 inches (4 to 5 cm) wide, and to whatever length is appropriate for the size mold that you want to make. Decide on the size of the sheet of paper that you want the mold to make and cut the slats accordingly, taking into account the thickness of the board. For example, a mold that makes a 9 x 12 inch (23 x 30.5 cm) sheet of paper should have side pieces 12 inches (30.5 cm) long and top and bottom pieces that are 10½ inches (26.5 cm) each—to account for the overlap of the thickness of the side pieces.

Use waterproof wood glue to join the pieces together, with a jig or temporary screws to hold the pieces in place. Once the glue has set, replace the screws with hardwood pegs: ¼ inch (6 mm) dowels or carved sticks work well. Drill out a peg hole with a ¼-inch (6 mm) drill bit, add a dab of waterproof wood glue, pound the pegs into place, and saw off the excess peg wood.

The ribs need to be flush with the top surface of the mold and secured with a ¼-inch (6 mm) dowel. Space the ribs no more than 1¼ inch (3 cm) apart from one another. Each rib should be slightly triangular in cross section to facilitate water shedding.

Thirty mesh (600 microns) brass screen (30 strands of wire per inch) is ideal for fine paper molds. Stretch and secure the screen onto the mold with brass or copper tacks so the mold doesn't rust and stain your paper. (This is known as "foxing.")

## MAKE A DECKLE

The deckle needs to fit the mold rather exactly, and therefore it should be made after the mold. The important thing is that the inner dimension of the deckle matches, or is slightly smaller than, that of the mold. This ensures easy transfer when couching the fresh waterleaf onto the felts.

## USING THE DECKLE WITH THE MOLD

Fit the deckle to the mold and draw a waterleaf from the pulp vat as you normally would. After the mold drains a bit, remove the deckle and proceed to couching onto the felts. The deckle prevents pulp build up on the margins of the mold and thus makes couching more consistent and successful.

Make more professional molds with ribs and brass screen secured with copper tacks. The mold on the left shows the ribbed supports as it awaits a new screen. Both have snugly fitted deckles with inner dimensions matching that of their molds.

# Make Pulp

Pulp is derived from plants whose bodies are largely made of fibers. These fibers are like the bones of the plants, and they're what provides paper with structure as well. The goal of the papermaker is to separate the fleshy material from the fibrils that make up the skeletal structure of the plant material. The process of pulp-making can have some variation, but it's generally accomplished through a series of cutting, pounding, boiling, and blending steps.

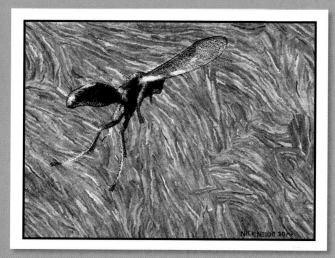

*The Original Papermakers.* Nick Neddo 2014, pine soot ink on wasp nest paper. Notice the parallel lines of color in the paper. Wasps gather fibers from plants, chew them into a pulp, and lay it down in exquisite, structural rows.

## MATERIALS

Cattail (*Typha* spp.) leaves
Water

## TOOLS

Hatchet or machete
Scissors
Bucket or enamel pot
4- to 5-gallon (15 to 19 L) enamel or stainless steel canning pot
Hot plate, stove top, or campfire
Filter materials, such as window screen, paint filter, or other fine mesh strainer
Blender
Two 5-gallon (19 L) buckets
Pint or quart (570 ml or 1 L) measuring cup
Colander
Freezer, optional

### 1: Find some fibrous raw material.

Anything that can be beat into a pulp is a good candidate for experimenting with making paper. Look for stringy, long-fiber plant materials, such as grasses, cattail, mosses, cedar bark, mulberry bark, basswood bark, milkweed, wild grape bark, ash bark, willow bark, iris, lawn clippings, corn stalks, rice stalks, bamboo sheathes, palm bark, cotton, and anything fibrous that piques your curiosity.

For the photo sequence in this chapter, I used common cattail (*Typha latifolia*) as a pulp source. Cattail marshes play a critical role in the ecosystem. In addition to pulp, they're an important source of food, medicine, shelter materials, weaving materials, and profound beauty. They also filter harmful pollutants from surface water.

### 2: Cut it into smaller bits.

Use scissors to snip lighter pulp materials into ½-inch (1.3 cm) or smaller pieces. To achieve this with heavier, harder materials, you may need to use a heftier tool, like a hatchet. Some material may need to be pounded with a wooden mallet or the back of a hatchet before it is ready to be made into ½-inch (1.3 cm) sections. If it is difficult to cut with scissors, pound it for a while and try again.

### 3: Boil.

Pour the snipped plant material into a 4- to 5-gallon (15 to 19 L) enamel or stainless steel canning pot until it is half full.

Next, add an equal amount of water, leaving a little space so it doesn't boil over. Use a hot plate, stove top, or campfire to boil the material for at least three hours. Check on the water level periodically, adding more if needed, to make sure you don't burn the batch.

## 4: Make pulp.

Set up your blender with your preboiled (but now at ambient temperature) plant material nearby, along with a clean 5-gallon (19 L) bucket, a pint or quart (570 ml or 1 L) measuring cup, and a supply of clean water. Use the measuring cup to scoop out a cup or so of plant material to put into the blender. Top it off with water and chop it up gradually, paying close attention to how well your blender deals with it. Most average blenders can handle this job if they are allowed to work with small batches at a time. Once the material is partially pulped, let the blender work on it uninterrupted for a minute or so before checking the consistency. Look for chunks of solid material within the pulp body. If this is present, blend it some more. If successive blending does not reduce the chunks into pulp, it may not have boiled long enough to pulp sufficiently. Repeat the boiling step and try again until all of the plant material is transformed into a lovely pulp.

## 5: Store.

If you're ready to make paper, you can pour your new batch of pulp into the paper vat. If you want to store the pulp for later use, you have some options. One option is to drain and freeze it. Use a paint strainer (from a paint or hardware store) or sheet of gauze or window screen to line a large colander, or bucket drilled with holes. Pour the pulp slurry through the filter to separate most of the water out. The resulting pulp wad can be bagged and stored in the freezer or actively dried out into a pulp paddy so it can be stored without the threat of molding.

**TIP:** If you smell a strange aroma or hear a struggling mechanical sound coming from your blender, turn it off. This is the smell of a blender on its way to the dump. Work with smaller batches of material at a time or consider retting the raw material.

Snip the pulp stuff (cattails, in this case) into 1-inch (2.5 cm) pieces and put them in a bucket or enamel pot. You may need to use a hatchet or machete to cut up more resistant pulp stuff.

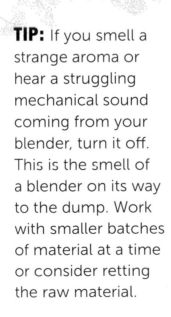

Boil the snipped plant pieces for at least three hours in a large enamel or stainless steel pot. This breaks the fleshy parts of the plant away from the fibrous skeletal parts. The substance of paper pulp is the result.

Blend the boiled plant material with water in small batches. Begin with course chopping before allowing it to blend for a minute or so. Go back and repeat the boiling step if the blender has a hard time with this.

## Retting with Lye (after step 3)

Retting is an optional step that speeds up the boiling process by facilitating the separation of the fibrils from the fleshy parts. You can accomplish retting by adding a small amount of lye to the pulp pot. Lye is a caustic chemical that can be made from the white ashes left over from burning hardwoods. It's simple to make, surprisingly useful, and relatively safe to use, provided you use a little caution. You can buy commercial lye in hardware stores, but making your own is more rewarding.

First, make a fire with hardwoods for fuel and enjoy it with loved ones or your own introspective ponderings. When the fire is out and the fire pit is cold, collect the white ashes that remain. Put enough in a 5-gallon (19 L) bucket to half fill it, keeping as much dirt, sand, charcoal, and pebbles out as possible. For folks who heat their homes with wood stoves, you need only to clean out and collect the ashes.

Fill the bucket nearly full with water and use a stick to stir the ashes into a thick slurry. Allow the concoction to sit for a while and then stir it again for a minute or so. Let it sit overnight, stirring it intermittently. The following day, pour the lye concoction through a strainer, directly into the enamel pulp pot with the plant material. Add more water if you need to bring the water line up, and then proceed to boiling.

With lye in the solution, the boiling time can be reduced to an hour or so. Afterward, the pulp material needs to be rinsed off to remove the lye, which will continue to break down the fibers if it is left with the material. Rinsing can happen outside on the lawn with a garden hose. Pour the pulp material and lye onto a porous sheet, like a window screen or sheet of burlap.

Spray it off with the hose until the material is no longer slippery to the touch. If you're not in a hurry you can leave it out to let the rain do this job for you. Another rinsing option is to use your bathtub and shower. Be sure to put a filter in the drain so you don't clog your plumbing!

When it's all rinsed out, you can proceed to pulping.

# The Paper-Making Process

The following instructions will give you a reliable procedure to follow for making homemade paper from your locally available pulp sources. You may encounter many variations as your apprenticeship with paper-making continues. My hope is that you gain confidence with the basic process outlined here, allowing you to experiment and take it in your own direction as your curiosity and creativity allow.

## MATERIALS

Pulp
Water

## TOOLS

Vat
Mold (and optional deckle)
Felts
Paper press

## 1: Fill the vat.

Pour 2 quarts (1.9 L) concentrated, wet pulp into the vat. Add 1 quart (1 L) warm water to the pulp and stir it all together with your hands. Getting the ratio of pulp-to-water just right takes some practice, and it depends on how thick or thin you want your sheet of paper to be. If the pulp slurry seems too thin and the fibers don't mat properly, add more pulp. If the slurry is thicker than you prefer, simply add more warm water until the consistency is more desirable.

The mold should fit into the vat, fully submerged at the bottom of the pulp slurry. If this is not the case, add 2 parts pulp to 1 part water (1 quart [1 L] pulp to ½ quart [500 ml] water) until the water line in the vat is high enough.

## 2: Form a waterleaf.

If you are using a deckle with your mold, fit it snugly over the screen side and hold them together as a unit. Stir the pulp in the vat so all of the fibers are in suspension within the water. Now hold the mold in a vertical position at one end of the vat. Use a slow and smooth scooping motion as you submerge one end of the mold into the vat while it is still vertical. As the mold enters the pulp slurry, level it out until it is horizontal and completely submerged by 3 or 4 inches (7.5 to 10 cm). In the same slow and fluid motion, lift the mold straight out of the pulp. When the mold and deckle have fully cleared the water line, give it a gentle shake

Reconstitute the pulp in the vat with 2 parts pulp to 1 part warm water. Add more pulp or water based on how thin or thick the slurry is. With a little experience, you will find a consistency that you prefer.

forward and backward as well as side to side. This slight agitation helps the pulp fibers interlace and is known as "throwing off the wave." Next, slant the mold to a 45° angle and let it drain for 10 seconds or so.

## 3: Couch the sheets.

When the mold stops dripping, it's ready to be couched (pronounced "cooch"), which is the action of transferring the newly formed waterleaf onto a felt blanket. Wet down your felt blankets and lay one out flat on your work surface. If you used a deckle, remove it and hold the mold horizontally with the waterleaf facing down, over the felt blanket. In one unbroken rolling motion and with evenly applied pressure, slowly press one edge of the mold down onto the felt. Apply pressure to the opposite side of the mold as you begin to lift up on the first edge. If done properly, the sheet will be transferred onto the felt blanket where it can be covered with another felt for the next waterleaf to be couched upon. A stack of waterleafs sandwiched between felt blankets is called a post.

This step can be tricky at first, but with practice you'll be able to couch sheet after sheet with consistency.

Hold the mold in a vertical position at one end of the vat. Slowly lower it into the pulp slurry with a scooping motion, until it is at the bottom of the vat, fully submerged and horizontal.

**2a**

In the same slow movement that you used to put the mold into the pulp, lift straight up out of the vat. Gently rock the mold side to side and front to back, and then tilt it to drain.

**2b**

**3**

Couch the waterleaf onto wet felt blankets by using a rolling motion from one side to the other. If the felts are not sufficiently wet, the fragile waterleaf is likely to be damaged in an unsuccessful transfer.

## 4: Dry the sheets.

If you don't want to bother with couching the sheets onto felt blankets, you can just leave the waterleaf on the mold and let it dry there. If you plan to make more paper that day, you will need to use that mold again, and couching will be necessary. To dry a post of waterleafs, you'll need to squeeze them in a press of some sort. Paper presses can be as simple as two flat pieces of wood with a heavy rock on top, or as complex as a hydraulic press that can deliver 15 tons of pressure per square inch (207 MPa). A bookbinder's screw press is ideal for paper-making, and you can make a modified version of one with two bamboo cutting boards and some simple hardware. (See "The Printmaking Press" on page 143.)

The purpose of pressing is to drain as much moisture from the paper as possible. Compress the post with whatever press mechanism you have. When water stops draining from the compressed post, remove it from the press. From here the paper will continue to dry on the felt blankets. Take the post out of the press and separate each felt, spreading them out on a flat surface. Allow the paper to dry over the next 2 to 7 days, depending on the ambient humidity in your work area. When the paper is dry to the touch, take the sheets off the felt blankets and stack them up again. Put the stack of paper back into the press for final drying. Expose the stack of paper to warm temperatures and/ or moving air to prevent it from molding in the press.

### TIPS:

If the felts come out of the press quite wet, consider adding dry felts into the post to help absorb the excess water. Sometimes you can safely transfer a damp waterleaf onto a dry felt and then press the post again.

"Dry to the touch" is somewhat subjective and is determined by several factors. How thick or thin the individual sheet of paper is, what kind of pulp material was used, and the relative experience of the papermaker all factor into the equation.

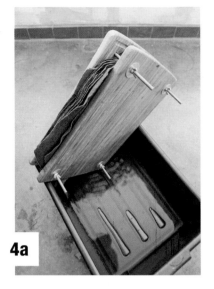

**4a**

Compress the post of felts and paper with a press of some sort. These can be made relatively cheaply and with few materials. Pressing removes the majority of excess water from the paper and gives it a smoother surface texture.

**4b**

Place finished sheets of paper onto the floor or hang to finish drying. When dry to the touch, return them into the press again to keep them flat.

**THE ORGANIC ARTIST**

# Sizing Paper

If you plan to use your handmade paper for watercolors or inks, you may want to size it first. Sized paper prevents the wet pigment from spreading and feathering beyond the intended area. New sheets of paper can be sized after they fully cure, which takes about two weeks after drying.

You will need a shallow pan or tray large enough for your paper to fit inside. Add 1 quart (1 L) warm water to the tray and have some newsprint paper (cut up brown paper bags work) at hand to spread out as a drying surface. Now dissolve one packet of gelatin (or use homemade hide glue and measure out 7 grams) in 1 cup (235 ml) of boiling water. When the gelatin is fully liquefied, add it to the quart (L) of warm water in the sizing tray and mix them together. This will be enough sizing to treat up to ten sheets of paper.

The goal is to fully submerge each sheet of paper into the sizing solution for a short time only, keeping it from getting completely saturated. Hold the paper on one end with both hands. Slowly submerge the sheet from one end, allow it to sit for a second or two, and then pull it out of the sizing solution from the opposite end. This ensures adequate coverage without letting the paper get soggy. Put the paper down on the prepared drying surface or hang to dry if space is limited. When the sized paper gets dry to the touch, consider squeezing it in the press again to keep it nice and flat as it finishes drying.

## EXPERIMENT

There is a lifetime of experimentation to be done in regards to making paper from the landscape. The possibilities are as varied as the number of plants in the world. Aside from making paper from pulp, the bark of some trees can be harvested and processed into a paper-like material for making art. Parchment is a paper-like material made from processed goat hides. Rawhide of other animals has been a surface to draw and paint on for millennia. Don't forget to think outside the box in your experiments, and most importantly, have fun.

*The Giant Forest*. Nick Neddo 2009, marker on birch bark. The bark of trees offer the creative and curious a lot to explore. You don't have to kill a tree just to get a panel of bark.

# 9 PRINTMAKING

In principal, printmaking is very simple. Complexities come about through the use of specific techniques, creative ambition, levels of craftsmanship, and material preparation.

Here is the basic printmaking routine.

1. Ink a found or prepared object.

2. Print said object onto another surface.

This chapter is intended to give you some options for making simple prints and printmaking tools, rather than being a detailed tutorial on how to make woodblock prints.

The tools and products of a simple printmaking studio, left to right: woodblock, bench hook, woodblock print, carving tools, wooden spoon, screw press, inking plate (palette), and brayer

# Tools and Materials for the Old Woodblock Printing Process

Just as with any art medium, the woodblock printmaking process depends on the use of special tools, some general, and some somewhat specialized.

The bench hook is a simple, yet important tool used to hold the woodblock in place while you carve it. They are easy to make even if you are not that handy with tools.

## THE BENCH HOOK

The bench hook is a simple tool that holds the woodblock in place while you work on it with your chisels. They are easy to make, and they're an essential part of the woodblock printmaker's studio.

They do not need to be fancy, just effective. With a bit of lumber, a saw, impact driver, and some screws you can make one in a morning and be using it by lunchtime.

## THE WOODBLOCK

Preferred woodblocks are made from planks (or plywood) of long-grained sections. This refers to the grain that is parallel to (follows) the growth of the trunk or branch that the block is cut from. Avoid pieces of wood that have knots or other inconsistencies. Woods of many species work well for block printing. In general, hardwoods allow for the clear transfer of fine details and hold up to the demands of the printing process better than softwoods. However, they're more difficult to carve into and require more frequent tool sharpening than softwood blocks.

## THE PALETTE OR INKING PLATE

The palette is used for mixing and applying ink to the brayer. Use a flat, smooth surface, such as a piece of glass from a picture frame; a shallow, glass baking pan; or the bench hook.

## BRAYERS

Brayers are used to apply ink onto the block, and they can be used to transfer the ink from the block to the paper. A paintbrush can also be used.

**TIP:** The bench hook can also be used as a palette. The harder and smoother the wood used, the more effective the surface will be as a palette. After sanding the surface to a smooth finish, oil the wood to keep it from absorbing the ink.

## PRINTING DEVICE

A simple screw press (also used in the paper-making process) can be used to make prints. You can also print by hand with the brayer or rolling pin or by burnishing the back of the paper with a wooden spoon. There are several tools for achieving a print for each culture that has a printmaking tradition.

## INK

Commercial printmakers' ink is available as oil or water based. Printmaking ink is distinctive from other inks because of its semi-thick, tacky consistency. This tackiness is helpful for applying ink to the block with the brayer. Thinner, less tacky inks can be applied to woodblocks with paintbrushes. You can convert your homemade inks for printmaking by thickening them with honey.

A simple mister spray bottle works well for stenciling objects with ink.

## Stenciling

Another technique for transferring images onto a surface is stenciling. Stenciling is almost the opposite of making a print, where everything but the stenciled object is pigmented. Human hands appear to be the original stenciled object, found on the walls of Paleolithic caves throughout the world. These ancient artists were probably mixing the pigments with water in their mouths and spray painting over their hands with a well-developed spitting technique.

These days, spray pump bottles are available for spritzing ink over stenciled objects onto paper. Cut-out shapes and found objects from the landscape offer endless possibilities for stenciled compositions. Experiment and have fun!

*Sit Spot*. Nick Neddo 2014, quill pen, walnut ink, and spray bottle with stenciled maple leaves

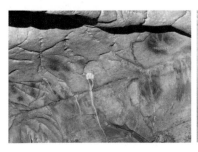 

32,000-year-old stenciled and printed hands in Chauvet Cave, France.

# The Rustic Brayer

For this project, you'll make an average-size brayer.

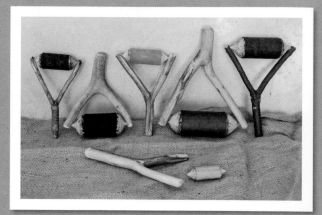

An assortment of handmade brayers gives you more creative options and control when inking the woodblock. Large ones provide efficient, general coverage. Smaller ones allow you to ink specific areas and detail work.

## MATERIALS

Y-shaped branch
Cylindrical stick
Oil, beeswax, or hard soap
Leather (1 square foot [30.5 cm²] is plenty)
Water
Strong thread

## TOOLS

Folding saw
Sharp knife
Pencil
Cutting board
Scissors
Leather needle
Thimble (optional)
Small container

## 1: Find a Y-stick.

The handle of the brayer is made from a tree branch that forks into a Y shape. Look for a tree that has fallen fairly recently. These fallen heroes offer the wildcrafter an abundance of material that's usually somewhat inaccessible. Choose a Y-stick that is between ½ and 1½ inches (1.3 and 4 cm) in diameter and close to symmetrical with a 5 inch (12.5 cm)-wide splay between the two forks. Use a folding saw to cut the handle end to about 4 inches (10 cm) long. Saw the forked ends to be as long as they need to be to have a 5-inch (12.5 cm) spread between the ends of the two forks.

## 2: Find a cylinder.

The cylinder is the roller mechanism for the brayer, and its level of symmetry directly corresponds to how smoothly it rolls. While you're on the prowl for the Y-stick, keep your eyes out for a straight, round stick, about 1 to 1½ inches (2.5 to 4 cm) in diameter. The trunks of saplings are usually good sources of sticks that are straight and close to perfectly round in cross-section. Like the Y-stick, it's best if the cylinder comes from a dead and dry tree so it doesn't shrink and fissure when it dries later. This can be from any species of wood as long as it is not cracked or punky (dry rotten).

**TIP:** Take your time when you're hunting for materials on the landscape. Be patient with your pursuit and keep the image of what you want in your mind. Think of it as a conversation with the landscape. With persistence, you'll find what you need for your project.

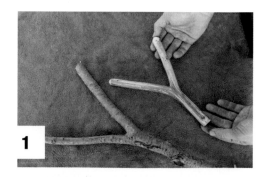

The Y-stick as a branch and as a refined brayer handle. Find Y-sticks that are more or less symmetrical in shape and in the diameter of each fork.

### 3: Debark.

Use the dull edge from the back of your knife to scrape the bark off of the cylinder. If this is not productive, you can whittle the bark off by using shallow slices. Be careful not to cut into the wood, making flat spots in your round stick. If, however, the stick needs to be made more perfectly round than it has presented itself to be, this is a good time to make those modifications. Again, use shallow slices when you whittle, staying in control of the knife and progressing toward more perfectly round gradually.

### 4: Whittle the ends of the cylinder.

Use a sharp knife to carve a point on each end of the cylinder. Make the points elongated as they become narrower and keep them centered in the middle of the cylinder's cross-section.

### 5: Make the handle sockets.

The sockets are where the pointed ends of the cylinder lock into the Y-stick handle. These are located at the ends of each fork and face each other. The distance between the sockets is less than the length of the cylinder by about ¼ to ½ inch (6 mm to 1.3 cm). This is important for locking the cylinder into the handle with the right amount of tension. Hold the cylinder up to the handle and position it so that the pointed ends overlap the forks of the Y-stick where they will fit

inside. These areas are where you will make the sockets and should be no less than 1 inch (2.5 cm) from the ends of each fork. Mark them with a pencil.

Hold the handle of the Y-stick to position it on a cutting board (or other hard surface) with one fork on the board and the other in the air directly above it. Next, place the tip of your knife along the pencil mark, in line with the upper fork. With light pressure begin to twist the knife in a drilling motion in both directions, gradually increasing downward pressure as needed until the sockets are ⅛ inch (3 mm) deep.

Invert the Y-stick to repeat this process on the other fork. Do your best to keep these sockets in the same plane and in alignment with one another. When both sockets match, lubricate them by putting some oil, beeswax, or hard soap into them. This helps the brayer roll with less friction.

### 6: Test the fit.

Place one pointed end of the cylinder into one of the sockets. With light pressure, gently flex the forks of the Y-stick apart only enough to snap the other point of the cylinder into the other socket. Now test the action by rolling it on a variety of surfaces. The cylinder should roll freely. After test driving your brayer, take it apart again so you can finish it.

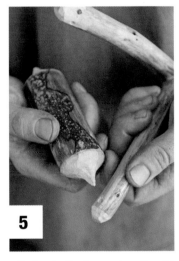

The brayer's roller: Find a stick that's as round as possible in cross-section, whittling it rounder if necessary. Notice that the whittled ends are more or less symmetrical to one another, with fairly long points.

Perforate the brayer handle with a hole near the end of each fork. Line these up in the same axis as much as possible and make them about ⅛ to ¼ inch (3 to 6 mm) deep.

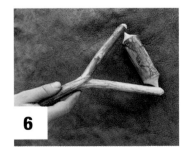

Fit the cylinder into the handle by inserting one end first. Gently flex the fork open and snap the other end in. Test the brayer and modify the pieces to get a fit with as little friction as possible.

**TIP:** If you have trouble assembling the brayer, you may need to reduce the length of each point slightly and try again. Other trouble-shooting may include decreasing the diameter of each point to match the socket, or making the sockets deeper and/or wider to accommodate the cylinder. Remember that the sockets and the points are custom made to match each other so make any necessary modifications to make them fit.

## 7: Prepare the leather.

The cylinder will be wrapped in leather to pad it and to allow it to absorb and transfer ink onto a woodblock. Thin leather works well because it is flexible and easier to sew than thicker leather. Use sharp scissors to cut a section of leather to the width of the cylinder, not including the length of the tapered ends. Cut this piece long enough that it can wrap around the cylinder, with each end of the leather cuts meeting. Don't worry about

making this a precision cut at this point, just make it long enough to make it around.

When this piece is cut out, soak it in a small container of water to thoroughly saturate it. After it's nice and wet, wring it out a little and spread it out flat on a cutting surface. Now is the time to make more precise measurements relative to the cylinder. Trim the width of the leather to match that of the cylinder, making clean and straight cuts. Wrap the leather tightly around the circumference of the cylinder and make a mark where the edge ends up. Trim these edges nice and straight so they make a clean seam when they're stitched.

## 8: Stitch the tube.

Fold the leather in half lengthwise so the grain side (the smoother textured side) is touching and the edges are lined up. This is a good time to check for any lack of symmetry and make adjustments. Thread a leather needle (often referred to as a glover's needle) with strong thread, such as upholstery or button thread, doubling it and knotting the two ends together. Begin on one of the corners of the lined-up seam and sew in about $\frac{1}{16}$ to $\frac{1}{8}$ inch (2 to 3 mm) from the edge. Thread the needle through the loop at the knotted end before pulling it snug. Make the next stitch $\frac{1}{16}$ to $\frac{1}{8}$ inch (2 to 3 mm) from the first, moving toward the other corner of the seam. Continue sewing the edges together, making your stitches as consistent as possible and tugging

the thread gently after each stitch to prevent slack in the line. When you get to the far corner of the seam, repeat the last stitch, knot the end securely into the stitch, and trim the lose thread.

## 9: Fit the leather pad onto the cylinder.

Turn the leather tube inside out, so the smooth textured surface is now facing out. Pull and tug at the seam until the stitched ridge flattens out. This will prevent the brayer from being too bumpy during use. Next, fit the leather tube onto the cylinder. This should happen when the leather is still wet, or at least damp. If it's somewhat difficult to fit the tube onto the cylinder you have done a good job, because it will be unlikely to fall off or move around in unintended ways during use. Push, pull, pinch, and scrunch the leather tube until it sits centered on the cylinder. Fit it onto the handle and let it dry. As the leather dries, it will shrink onto the cylinder, making an even tighter fit.

If it's easy to fit the tube onto the cylinder, it may be too lose when it dries. If this is the case, set the first tube aside, save it for a larger diameter cylinder, and make another tube with a tighter fit.

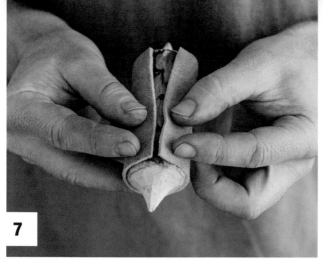

**7**

Cut a rectangle of leather to the size needed for it to fully wrap around the cylinder. Soak it in water, stretch it out, measure, and trim again. Cut it slightly short in length to ensure a tight fit.

**8**

Fold the wet leather in half lengthwise, smooth-side-in and stitch the ends together with a whip stitch. Make your stitches tight and even.

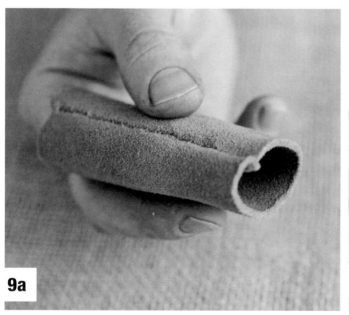

**9a**

After stitching, turn the leather tube smooth side out and stretch the seam until it lays flat.

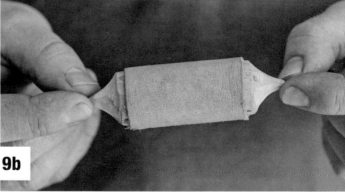

**9b**

Fit the leather tube over the cylinder while it is still wet. This should be somewhat difficult, requiring persistence in squeezing, scrunching, pinching, and pulling to join the two. As the leather dries, it will shrink tight around the cylinder.

**TIP:** You may want to have a thimble or thick piece of leather on hand to help penetrate the needle through the leather.

The finished brayer in use to ink a woodblock

## Using the Brayer

You can use the brayer to apply ink to a woodblock and to transfer the ink from the block to the paper.

To ink a woodblock, you first have to load the brayer with ink, using a palette (ink tray). Apply the ink to the palette and roll the brayer across it from every direction until the brayer has a consistent density of ink, but is not saturated. The goal is to find the balance of having enough ink to print cleanly, but not so much that the fine lines of the woodblock get clogged. With experience you will learn the unique traits and capabilities of each handmade brayer.

To transfer the ink from the block to the paper, a clean brayer is rolled across the back of the paper as it lies against the block.

**TIP:** When loading the brayer, listen for a sound that resembles that of bacon frying as an indicator that it has gathered the right balance of ink.

## Rubbings

Rubbings are simple ways to "print" an image onto a piece of paper. Crayons and charcoal are great mediums to make rubbings with. Find a surface with interesting texture, such as a tree trunk or boulder, and place a sheet of paper on it. Thin paper will transfer the more subtle details of the texture underneath, but must be used with a delicate approach. Use the lengthwise edge of the charcoal or crayon to record the texture underneath as you gently rub it across the surface that you wish to be recorded. You can gather smaller objects and place them onto a flat work surface to rub on top of. The possibilities are endless.

# OPTIONS FOR PRINTING

Printmaking is a vast and dynamic artistic medium, partly because of the many options available to accomplish the print.

## The Printmaking Press

Simple presses can be devised for printmaking using two flat boards, some felt blankets, and some weights. The amount of weight used to successfully transfer the ink from the block to the moist paper is a variable you'll have to play with for your situation.

## The Simple Screw Press

Another option for a press is a simple screw press. Make one with two bamboo or hardwood cutting boards, four carriage bolts with two washers, and a butterfly nut for each. Clamp the boards together and drill a hole straight through at each corner. Slide a washer onto each bolt before inserting them through the holes. A second washer is used for each bolt on the other end before the butterfly nut is screwed on. This press can be used for years of service, and it will come in handy for a variety of tasks.

The brayer can be used to transfer the ink onto the paper by rolling it thoroughly (perhaps excessively) onto the back of the paper as it lays on the block. Choose only the most perfect brayers with smooth, fluid action for this. If there is any imperfections in the shape of the cylinder, you will most likely not be satisfied with the printing results.

Printing presses can be simple. The essential elements are two flat boards cushioned with felt blankets and a way to apply pressure, either by weight or clamping down.

A fresh block print made using a wooden spoon

The back of a wooden spoon works well for transferring prints from the block onto paper. Rub the back of the paper in a burnishing motion. With practice, this method can give you a satisfying level of control in printing.

# 10

# SKETCH-BOOKS AND JOURNALS

An assortment of the author's sketchbook journals made with coptic stitched binding. Cover materials include: cow rawhide, pin cherry bark, white birch bark, salvaged book cover with handmade paper glued on, cardboard, and thick handmade paper

The art of bookmaking encompasses countless techniques and approaches to accomplish a fairly simple goal: to tie a stack of pages together. Making a book is essentially that simple, but the number of techniques and level of sophistication possible in bookbinding is as varied as art itself. Each style of binding has a unique set of aesthetic and functional characteristics. With these factors in mind, I would like to introduce you to one of my favorite bookbinding techniques.

# The Coptic Stitch

This method of bookbinding is one of my favorite options for making sketchbooks and journals. Coptic stitch–bound books have the ability to open and lay flat, which is a trait that is appreciated by artists and writers alike. Other binding options may be more approachable, but they lack this important feature.

For this project you will use a sketchpad to learn the basics of bookmaking. After learning the basics here, you can experiment with your own handmade paper and book covers.

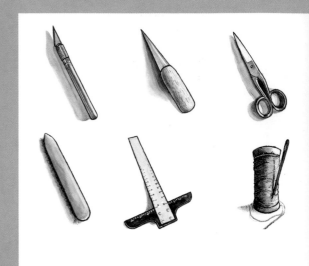

Tools used in this chapter: utility knife, awl, scissors, bone folder, T-square ruler, and long needle

## MATERIALS

9 x 12 inch (23 x 30.5 cm)
   sketchpad
Strong thread
Beeswax

## TOOLS

Bone folder (optional)
T-square ruler
Pencil
Utility knife
Awl
Scissors
Book press (optional)
Large needle

## 1: Make signatures.

A signature is a group of four sheets of paper that are folded in half and nested into one another. Each signature makes 16 pages of a book. Tear out the pages from the sketchpad and fold one exactly in half by lining up the corners before making the crease. Serious bookbinders use a tool called a bone folder to make a tight crease, but if you don't have one, you can use your thumbnail instead. Repeat this with the next three sheets of paper, nesting them into each other as you go until you have your first signature. Make eight signatures in all.

## 2: Cut backing board into cover pieces.

The rigid board that backs the sketchpad is the book cover. Use a T-square ruler, a pencil, and a utility knife to measure, mark, and cut this in half to make two 9 x 6-inch (23 x 15 cm) pieces.

## 3: Measure and mark.

Set aside one of your signatures and mark the crease with a pencil at 1, 2, and 3 inches (2.5, 5, and 7.5 cm) from each end, making six marks in all. This marked signature will be a template for marking the other ones. Put this signature back into the stack with the others and keep them all in tight alignment with one another. Use a cover piece's straight edge to draw a line on the spine of your signature stack at each pencil mark, extending the marks to the other signatures with each vertical line.

Then, on one of your cover pieces, begin about a ½ inch (1.3 cm) in from the spine edge and use the same template to mark it as well. The cover should also have six marks measured at 1, 2, and 3 inches (2.5, 5, and 7.5 cm) from both ends.

## 4: Make holes.

At each mark on the cover piece, use an awl to make holes big enough for your needle to fit through. The first cover piece is now a template for marking the second piece. Line up the covers and puncture the second cover by going through the holes on the first.

Next, pierce holes in each of the signatures where they are marked. One at a time, flatten each signature out and make a hole straight though the four sheets of paper at each measured mark.

**TIP:** Make sure the sheets of paper are in perfect alignment when you open them up to make holes.

**1**

Fold each sheet of paper exactly in half and nest four of these together to make a signature.

**2**

Cut the cover board in half to make the front and back cover pieces for your book.

**3a**

Use a marked signature as a template to mark the cover. These should be about ½ to 1 inch (1.3 to 2.5 cm) from the cover's edge.

**3b**

Use the straight edge of the cover to mark all of the signatures at once, with the measured and marked one as a template at the top of the stack.

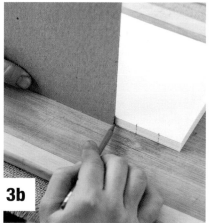

**4**

After measuring and marking the signatures and book cover material, pierce holes through them with an awl.

## 5: Sew the cover and first signature.

Thread the needle with waxed linen, cotton, hemp, or other thread and tie a knot at the end. If it is not waxed, run it along a chunk of beeswax until it is coated. Make sure you have a fair amount of length, but not so much that it is cumbersome to work with. It's not a big deal if the thread runs out before you finish.

Bind the signatures in the order that you marked them (to keep them in better alignment) and begin with the first one and the front cover piece. Sew from the inside of the signature, to the outside, then around to the outside of the cover. Then, loop around the stitch between the signature and cover and back into the signature (this time from the outside, in). Repeat this process for the next five holes on the cover, progressing from hole to hole on the inside of each signature. For the last hole of the signature, loop around and then into the first hole of the next signature.

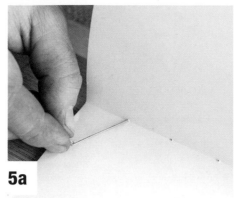

**5a**

Sew the cover with the first signature. Begin on one end by threading through the signature hole from the inside, along the crease.

**5b**

Now go around and come up through the cover hole on the outside of the cover. The needle will come out between the cover and signature.

**5c**

Loop around the stitch that connects the signature to the cover, pulling snugly.

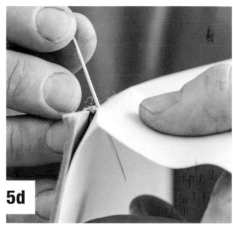

**5d**

Now go back into the signature hole and tug the thread gently to secure and tighten the knot.

**TIPS:**

Pull the thread slowly to minimize tangling and unintentional knotting.

After looping around each stitch, tug gently on the string to keep the stitching tight.

Continue to the next hole and repeat the pattern. When you come to the last hole of the signature, continue as you were, but this time sew into the first hole of the second signature after looping around the stitch.

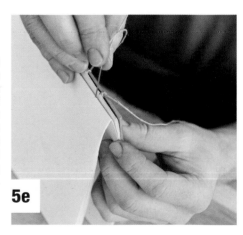

**5e**

## 6: Stitch the middle signatures.

Continue this same stitching pattern with all (except the last) of the remaining signatures, always looping around the stitch from the previous signature before entering through the signature holes. When you reach the last hole of each signature, enter the first hole of the next signature after looping around the stitch from the previous one.

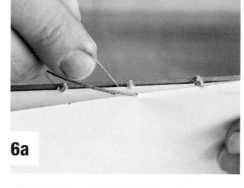

**6a**

Repeat the pattern, but this time loop around the stitch from the previous signature before sewing back into the current one.

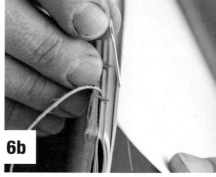

**6b**

Last hole: Loop around previous stitch and enter first hole of third signature. The third signature's stitches get looped around that of the second signature. Repeat this pattern with each additional signature, always looping around the previous one's stitches.

## 7: Stitch the last signature and the cover.

When you only have one signature left, it is time to put on the other half of the cover. Begin with the cover, stitching from the outside and exiting between the cover and last signature. Loop the needle around once, then into the signature. For the rest of the signatures, loop around the previous signature's stitch first, then through the cover (from the outside, in). Loop around the stitch from the last signature once, then back into the signature.

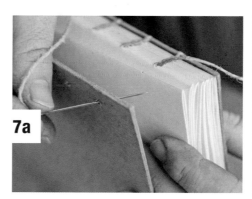

**7a**

Stitch the back cover and the last signature at the same time. Go through the cover first.

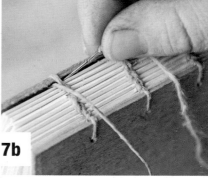

**7b**

Then loop around once . . .

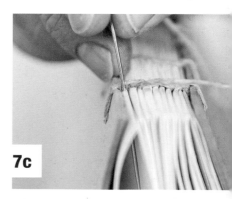

**7c**

Enter the signature . . .

**TIP:** When looping around stitches, be consistent with which way you make the loops: either all clockwise, or all counter-clockwise. This will make a more uniform pattern on the spine of your book. Also, to help you get a sense of the pattern, keep in mind that each stitch between the signatures gets "looped around."

## 8: Finish.

When you enter the last hole of the last signature, it's time to finish the binding. Tie off the lose end to the adjacent stitch nestled in the crease of the signature. A simple overhand knot or surgeon's knot works well. Tuck the extra length in under the previous stitch and snip off the extra thread. You have just made a sketchbook journal with coptic-stitch binding. Nice work!

**7d**

Come out the 2nd signature hole and loop around the stitch of the previous signature, as usual . . .

**7e**

Go through the cover, from the outside . . .

**7f**

Loop around the cover stitch . . .

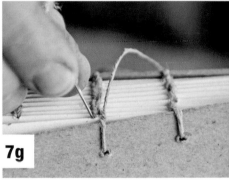

**7g**

And then back into the signature. Do this for the rest of the signature holes.

**8**

Once back inside the last signature, tie the remaining thread off to the stitching within the crease. Snip the excess thread, and you're finished!

# BOOK COVER OPTIONS

You can make hard or soft cover books, depending on what you want and/or what kind of materials you can come up with. In the salvaged or repurposed category, you have a lot of options. Cardboard and paperboard are abundant on the modern landscape and are more valuable resource than we may be aware of. Glue two pieces of cardboard together with the corrugation grain perpendicular to one another to make it more rigid. Keep this in a book press to compress the layers of cardboard together tightly as the glue dries.

The covers of old and discarded books are perfectly suited to continue being book covers. Consider gluing some fabric or nice paper over the old cover to reclaim it for your project.

Thin layers of bark from young trees can be glued onto rigid material to make an attractive laminated book cover. Other options to consider are rawhide, leather, and thin slabs of wood or even thick homemade paper.

# Gallery

*Female Northern Parula with Nest*
Nick Neddo 2014, paintbrush and pine soot ink.

*Hummingbird and Black Bear*
Nick Neddo 2014, paintbrush and black walnut ink.

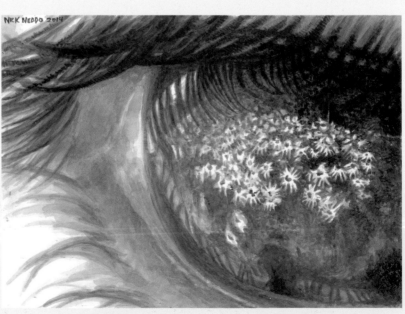

*Woodland Waterfall*
Nick Neddo 2014, paintbrush and pine soot ink.

*In The Eye of the Beholder*
Nick Neddo 2014, paintbrush and pine soot ink.

*Coyote Talking in the Grand Tetons*
Nick Neddo 2014, paint brush and acorn ink.

*Boy Meets Prairie Dog*
Nick Neddo 2014, paintbrush and pine soot ink.

# Bibliography

Cameron, Alison Stilwell, *Chinese Painting Techniques*, New York, Dover Publications, Inc. (1968)

Campell, Paul Douglas, *Earth Pigments and Paint of the California Indians*, California, Asia Pacific Offset (2007)

Dutton, Denis, *The Art Instinct*, New York, Bloomsbury Press (2009)

Fuga, Antonella, *Artists' Techniques and Materials*, Trans. Frongia, Rossana M. Giammanco, Los Angeles, Getty Publications (2006)

Gair, Angela, *The Artist's Handbook*, New York, Barnes & Noble Books (2002)

Harrison, Lorraine, *Artist's Materials*, Canada, Firefly Books (2005)

Heller, Jules, *Paper-Making*, New York, Watson-Guptill Publications (1997)

Johnson, Arthur W, *The Practical Guide to Craft Bookbinding*, London, Thames and Hudson Ltd. (1990)

LaPlantz, Shereen, *The Art and Craft of Handmade Books*, New York, Lark Books (2001)

---. *Cover to Cover: Creative Techniques for Making Beautiful Books*, North Carolina, Lark Books (1995)

LaFerla, Jane, *The Pendland Book of Handmade Books: Master Classes in Bookmaking Techniques*, New York, Lark Books (2004)

Okamoto, Naomi, *Japanese Ink Painting: The Art of Sumi-e*, New York, Sterling Publishing Company, Inc. (1996)

Peot, Margaret, *Make Your Mark*, San Franscisco, CA, Chronicle Books LLC (2004)

Saddington, Marianne, *Making Your Own Paper*, Pownal, Vermont, New Holland Publishers Ltd. (1992)

Saito, Ryukyu, *Japanese Ink Painting*, Vermont, Tuttle Publishing (1959)

Smith, Ray, *The Artist's Handbook*, New York, Alfred A Knofp (1997)

Studley, Vance, *Make Your Own Artist's Tools And Materials*, New York, Dover Publications, Inc (1979)

Turnes, Jane, *The Dictionary of Art*, London, MacMillan Publishers Limited (1996)

Watson, Aldren A, *Hand Bookbinding*, New York, Dover Publications, Inc. (1996)

---. *The Art and Craft of Handmade Paper*, New York, Dover Publications, Inc. (1977)

# Index

# Acknowledgments

A number of people have offered their support to help this book come to fruition.

Thanks to my parents, Dave and Deb, for letting me go feral as a child in the boundless wilderness of central Vermont. They encouraged us kids to use our imagination, think critically to solve problems, and make art. They always found a way to support my exploration of the arts, even when times were tight. Their support undoubtedly helped my path into adulthood unfold to the extent that it has.

Huge thanks and bottomless gratitude to Andy and Carolyn Shapiro for their sage advice, unwavering generosity, enthusiastic affirmations, studio space, and delicious meals. It's hard to imagine how I would have pulled it off without your help. Thanks to Mike Kinnerson for his persistent friendship, technical woodworking expertise, love of nature, and willingness to let me make a mess in his workshop. Thanks to Brad Salon and Sarah Corrigan of ROOTS School, for sharing a path less traveled in the pursuit of reconnecting people to innate genius and the living world around us. Also for offering great research resources and genuine enthusiasm and being the best co-instructors in wilderness living skills and primitive technologies that I have had the pleasure to work with. Thanks to Tim McFarlane of McFarlane Apiaries for his friendship, loving stewardship of bees, and donations of high quality beeswax, honey, and inspiring images.

Thanks to Susan Teare, who took the pictures for this book. I couldn't have anticipated how much fun it would be to work with you and I couldn't imagine being matched up with a more talented and easy-going photographer. Big thanks to my editor, Jonathan Simcosky, for hunting me down and pitching the idea for this book. Without him, the contents of this book would have remained isolated to my personal studio and the workshop format.

The biggest thanks and gratitude goes to my sweetheart, Sarah Shapiro. The creation of this book took over our home and time together in no insignificant way. Without her patience, resilience, aesthetic perspective, and proofreading, as well as financial and caloric support, this project would not have been possible.

# About the Author

Nick Neddo is a sixth generation Vermonter who has been making art since he could first pick up a crayon. He grew up exploring the wetlands, forests, and fields of his bioregion and developed a profound curiosity, respect, and love for the community of life around him. As a young teenager, Nick identified primary focuses that would become life-long pursuits: study of the natural world, Stone Age technology (popularly known as primitive skills), and creating art. Trusting the inherent value of these skills, he continues to embrace their pursuit with a ravenous appetite fueled by a genuine love of the living world and the creative process. He has traveled the country extensively, visiting the last great wildernesses, seeking traditional skills, and experiencing the landscape's majesty, which are common themes in his artwork.

Nick has been teaching wilderness survival and living skills, tracking, drawing, and nature awareness professionally since 2000, although he considers himself a perpetual student. He currently instructs at ROOTS School in Vermont, as well as other venues. You can find his latest artwork and other creations at www.nickneddo.com.

# Also Available

The Joy of Foraging
978-1-59253-775-4

Green Guide for Artists
978-1-59253-518-7

Adventures in Bookbinding
978-1-59253-687-1

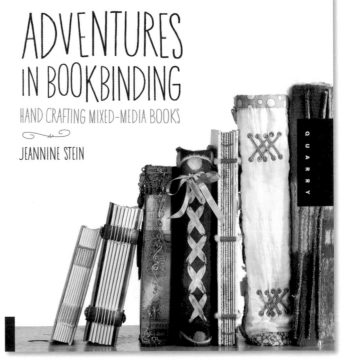